ETHER DAY

ETHER DAY

The Strange Tale of America's Greatest Medical Discovery and the Haunted Men Who Made It

JULIE M. FENSTER

HarperCollins*Publishers*

HarperCollins books may be purchased for educational, business, or sales promotional use. For information, please write: Special Markets Department, HarperCollins Publishers Inc., 10 East 53rd Street, New York, NY 10022.

FIRST EDITION

Designed by Joseph Rutt

Printed on acid-free paper

Library of Congress Cataloging-in-Publication Data
Fenster, J. M. (Julie M.)
Ether day : the strange tale of America's greatest medical discovery and the haunted men who made it / Julie M. Fenster
p. cm.
Includes bibliographical references and index.
ISBN 0-06-019523-1 (alk. paper)
1. Anesthesia—United States—History—19th century.
2. Ether (Anesthetic)—United States—History—19th century.
I. Title.
RD80.3.F46 2001
617.9'6'097309034—dc21 00-054117

01 02 03 04 05 ❖/RRD 10 9 8 7 6 5 4 3 2

Ether Day is dedicated to the libraries of Boston, including:

Boston Athenæum

Boston Public Library

Francis A. Countway Library of Medicine, Harvard Medical School

Houghton Library of Harvard University

Massachusetts Historical Society

CONTENTS

ILLUSTRATIONS

PROLOGUE

THE LAUGHING-GAS JOKE

In 1845 the New York *Daily Tribune* published a detailed account of an amputation. The operation took place at New York Hospital, a five-acre nest of low brick buildings, located on what is now Lower Broadway. The patient was a young man, cradled tenderly the whole time by his father and at the same time held firmly—and brusquely—in place by the attendants. As the surgeons—there were two—made their cuts, the boy's screams were so full of misery that everyone who could left the room. The first part of the operation complete, the young man watched "with glazed agony" as the chief surgeon pushed a saw past the sliced muscles, still twitching, and listened as the blade cut through the bone in three heavy passes, back and forth. That was the only noise in the room, for the boy had stopped screaming.

A scant four blocks up the street at the cavernous Broadway Tabernacle theater, close to four thousand people were howling with laughter at the antics of people who had inhaled nitrous oxide gas in the "Grand Exhibition" staged for just that purpose.

The audience's favorite was the fellow who set upon the master of ceremonies, Gardner Q. Colton, with his fists. He had to be dragged off by the "twelve stout men engaged," as the playbill explained, "to prevent those who take the Gas, while under its influence, from injuring themselves or others." Under the influence of nitrous oxide, volunteers were known to be insensible to pain, doing things that hurt only after they came to.

The critic for the highbrow *New Mirror*, writing a column called "Diary of Town Trifles," especially liked the "young man who coolly undertook a promenade over the close-packed heads of the audience."

"The impertinence of the idea seemed to me in the highest degree brilliant and delightful," applauded the critic, who then described other onstage escapades under the influence of laughing gas: "One silly youth went to and fro, smirking and bowing, another did a scene of 'Richard the Third,' and a tall, good-looking man laughed heartily, and suddenly stopped and demanded of the audience, in indignant rage, what they were laughing at!"

They were laughing at everything. As far as the audience was concerned, that was the whole idea of the "Grand Exhibition of the Effect Produced by inhaling Nitrous Oxide or EXHILARATING or LAUGHING GAS!" A few of those present at the Tabernacle temporarily lost their senses from inhaling the gas, but hundreds in the audience lost their senses from laughing at the antics onstage. Some people walked across heads, and some people had their heads walked on.

"This is in truth a scientific entertainment, although we have our doubts whether the use of it will serve to advance the cause of science," stated the *Tribune's* report on the nitrous oxide demonstration. But the *Tribune* was off—off by a scant four blocks in some uptown version of glazed agony. Any one of the four thousand people at the Broadway Tabernacle on that day in 1845 could have made the discovery that nitrous oxide was, on the contrary, at the core of the very greatest advance that "the cause of science" could receive. A couple of whiffs would have helped separate that boy at New York Hospital from the pain of surgery.

His operation was like any other between the beginning of time and Ether Day or, more pointedly, between the middle of

1800 and October 16, 1846, an era when help that was desperately required in hospitals was dismissed as a joke called laughing gas. During the same years, nitrous oxide was joined in the world of young fun by ether, a chemical with similar powers. Each one could make people insensible long enough for an operation, but ether was the stronger of the two. The laughing gas joke did not begin with the discovery of these substances. It began in 1800, when the English chemist Humphry Davy stated straight out in a well-received book: "As nitrous oxide in its extensive operation appears capable of destroying physical pain, it may probably be used with advantage during surgical operations."

No one took his suggestion, however. That was the laughing-gas joke. For almost fifty years, all over the United States and all over the world, the art of surgery was stranded on the screams of its patients. In an operating room in New York a young boy fell silent from the horror, and in some other place the brilliant Humphry Davy was duly knighted by the Crown and idolized by commoners, but meanwhile, the wisest thing he ever said was ignored, amid all the laughter that followed nitrous oxide and ether around.

1

ETHER DAY

On Friday, October 16, 1846, only one operation was scheduled at Massachusetts General Hospital.

That was not unusual: At Mass General, the third most active center of surgery in the country, operations averaged only about two per week from the 1820s through the mid–1840s. Operations were special events in that era—a long ordeal of an era that was to end on that very day.

The patient, a housepainter named Gilbert Abbott, had been admitted to Mass General earlier that week, expecting to have a large growth cut from the side of his neck. It was the most unremarkable of cases, except that its very predictability suited it to another purpose entirely. With the Gilbert Abbott case, the chief surgeon at the most renowned hospital in New England decided to give William T. G. Morton—a twenty-seven-year-old dentist whose salient characteristic was an excess of charm—a patient on whom to demonstrate a secret compound that promised painless surgery.

The day before the operation was to take place, Morton had received a letter from the hospital: a letter he had been hoping to receive.

"Dear Sir," it began, "I write at the request of Dr. J. C. Warren to invite you to be present on Friday morning at ten o'clock, to administer to a patient, then to be operated on, the preparation which you have invented to diminish the sensibility to pain." The letter was signed by the house surgeon at Mass General, C. F. Heywood, but the invitation had come from the hospital's chief of surgery, John Collins Warren, the veritable dean of surgery in the United States.

Morton's immediate reaction was fear—panic, in fact—that pain wouldn't be the only thing he'd kill on Friday morning. He had tested his preparation on a few dozen of his dental patients during tooth extractions, but he didn't know whether the same amount would be sufficient in a medical operation, whether more of it would kill the patient, or whether the apparatus he used had any flaw that might allow for an accidental poisoning. In fact Dr. Morton didn't know much as he held Heywood's note in his hands, except that Dr. John C. Warren would be watching his every move the next morning at ten.

And Dr. John C. Warren's salient characteristic was an utter absence of charm.

Morton's secret concoction was made of exactly two ingredients: first, sulfuric ether, a common liquid compound with a sweet pungency; and second, oil of orange to disguise the smell of sulfuric ether—that's what made it a secret, and that's what made it Morton's. There are other compounds known as "ethers," such as chloric ether (a rather sinister cousin to the sulfuric form). However, unadorned with a prefix, "ether" refers to sulfuric ether.

In early experiments Morton's dental patients had inhaled ether fumes from a cloth doused with the liquid. Morton soon replaced the cloth with a more elaborate inhaling apparatus that gave him greater control in two ways: over the delivery of fumes to the patient and, more to the point, over the commercial potential of his discovery in the medical world. In neither case did William Morton quite understand the thing that he was trying to control, but he had his apparatus with which to attack them both, and with it under his arm he went rushing into the street upon receiving his letter from Mass General.

Morton's destination was the workroom of Joseph M. Wightman, a specialist in the manufacture of scientific instruments. A small industry in making scientific instruments had been launched in Boston about fifteen years before, in the early 1830s, with the arrival of Josiah Holbrook, an enthusiastic Yale graduate whose goal was to introduce science to the general populace. Idealistic but utterly practical, too, Holbrook recognized that science in any form requires paraphernalia—great, endless closets full of *stuff*—if it is to maintain a bridge back and forth between abstraction and actuality. Yet scientific equipment was prohibitively expensive when Josiah Holbrook arrived in Boston. Harvard University boasted whole arrays of the latest apparatus, but only a few hundred people had access to any of it.

Holbrook's aim was to bring the new sciences to the American people, not merely to its professors. A biographer summarized his achievements in that regard in a single sentence: "An orrery [a model of the solar system], for example, similar to the one for which Harvard had paid $5,000 in Paris, Holbrook manufactured and sold for ten dollars." Holbrook was not the only man in the United States promoting science so widely, but on his arrival in Boston, he was among the very few. By the 1840s science education was a small industry, centered in Boston, and without it, a man such as William Morton might have had nowhere to turn for help in a rush on the day before Ether Day.

Even in repose William Morton could give the impression of a man in a rush, with his attention hopping from subject to subject. Where others had poise, he had energy. Morton was a strikingly attractive man, according to those who saw him in person. And judging by photographs and portraits, he did make a flashy impression, much like that of a leading man on the stage. He was well-proportioned, being on the tall side with a medium build that people were inclined to think was elegant. Morton had dark, wavy hair and intense blue eyes, features that were set off throughout most of his life by an extravagant moustache. He dressed extravagantly, too, in rich fabrics when plain ones were in style, loud silk scarves flowing out from under his lapels, and ornate buttons punctuating the cut of his coats. Overly optimistic and then pessimistic by turns, Morton dominated situations simply by presenting everything that had happened so much more starkly than did anyone else. To listen to him was either to be bored silly by all the manufactured drama or to be taken along for a wild ride. Nearly everyone who met William Morton seemed to fall in for the ride, at least for a time.

The drama of October 15 was real, with the operation at Mass General looming the next day. Nonetheless, in Morton's hands it took on an urgency or even a frenzy.

In the otherwise serious world of Boston medicine, only William Morton would have been so ill prepared for an appointment as important as his with J. C. Warren—and only he would have worked so hard to make it even worse. All in a hurry at midday on Thursday, October 15, Morton arrived at the workrooms of Joseph Wightman, a former Holbrook assistant. As of that day, the best thinking of Wightman and Morton on the subject of delivering ether fumes to a patient consisted of a glass globe containing a sponge soaked with liquid ether. A long stem on one side of the globe (or "retort," as the glass sphere was known), went to the patient's mouth. The same opening also had to let air into the globe, in order to keep the patient from suffocating due to lack of oxygen or from sustaining lung damage from the effects of ether fumes at full strength.

"He called upon me in great haste," Wightman later wrote of William Morton's visit that Thursday afternoon,

> and begged me to assist him to prepare an apparatus with which he could administer the Ether to a patient at the hospital the next day, as Dr. Warren had consented to use it in an operation. He appeared much excited; and although, from a pressure of other engagements, it was very inconvenient for me, yet I consented to arrange a temporary apparatus under these circumstances.

Wightman reconfigured the stopper at the end of the stem, "having a cork fitted into it instead of a glass stopper, through which cork a pipette or dropping tube was inserted to supply the Ether as it was evaporated. I then cut several large grooves around the cork to admit the air freely into the globe to mix with the vapor and delivered it to Dr. Morton."

As the hours grew late on Thursday night, though, Morton was still unsure of his chances of success the next day, and he decided to consult with one of Boston's most respected, and approachable, physician-scientists, Augustus Addison Gould. That would not have been especially difficult, since Morton was boarding in his house at the time.

Several months earlier Morton had used a hearty recommendation from his mentor, Charles Jackson, in order to obtain an appointment with Gould on the pretext of requesting permission to use Gould's name as an endorsement in his dental advertising. He received that and, within weeks, an invitation to board with Gould and his wife in their home on Boston's swank Colonnade Row: a line of glistening town houses designed by the brilliant Charles Bulfinch.

Augustus Gould had not been born to wealth; he'd been poor most of his life. That may have been one reason for his warm feelings toward William Morton, who had also been raised in straitened circumstances. In Gould's era earning a medical degree was no guarantee of a great or even a good living; however, he had become one of Boston's busiest practitioners, specializing in what would today be considered internal medicine. With the sort of duality that was common among Boston's medical men, he was also involved in another of the scientific specialties evolving at the time. Gould's avocation was conchology, or the study of seashells and invertebrate animals. He came by his interest as a poor boy might: wandering along the shore near his home in New Hampshire. He eventually had five shells named after him, including one with the felicitous name Gould's Bubble—*Bulla gouldiana.*

"The evening previous," Dr. Gould would recall in his account of the October 16 operation,

> Dr. Morton called to ascertain about the probable injurious effects of ether, and what articles might be used. I

> answered; and in the course of the conversation I asked him how he gave it. He told me that he put a sponge in a globe saturated with ether, and drew the vapors through a tube attached; breathed out and in through a tube attached. I suggested that the application of valves, to prevent breathing back the air into the globe, would be desirable, and sketched a plan. He said, "that is it; that is just it. I will have it for tomorrow."

"I advised him not to attempt it," Dr. Gould continued, "but to use what he was sure he would succeed with. He then left me."

The trouble with Dr. Gould's good advice was that there was nothing in the world that William Morton *was* sure he would succeed with. Other people were keenly aware of that fact. They advised Morton—they begged him—not to attempt anything at all. The same friends also turned to his wife, Elizabeth. "The strongest influences had been brought to bear upon me to dissuade him from making this attempt," she said later. "I had been told that one of two things was sure to happen: either the test would fail and my husband would be ruined by the world's ridicule, or he would kill the patient and be tried for manslaughter."

Nonetheless Elizabeth stayed up late at the Goulds' home, helping as William tried to work out a new design for his apparatus, one with valves, as Dr. Gould suggested, to stop the reflux of exhaled air.

Opting not to depend any further on Joseph Wightman's good nature, or what was left of it after the rush job the previous afternoon, Morton took his inhaler to another of Boston's scientific manufacturers. "I rose at daybreak," Morton said, "went to Mr. Chamberlain, an instrument-maker, and by great urging, got the apparatus done." After examining both the apparatus and Gould's new sketch, which Morton had taken care to bring, Chamberlain agreed to try to fit two valves onto the tube leading from the retort, to allow a patient to inhale from the globe and then exhale into the room. He worked on it as quickly as he could and maybe even more than that: as quickly as Morton could make him go.

Morton must have realized that Dr. Gould had been right. It would have been better to go into the operation with an inhaler known to function properly. That fact should have been obvious. As it was, he watched the time drawing nearer and nearer to ten o'olock, and he had no inhaler at all, just a pile of parts on Chamberlain's bench. Even if it could be finished in time, he would have no time to familiarize himself with the working of the valves, no chance at all to test it. If he had a moment that morning to consider his regrets, he must have wished he had stayed in bed and allowed ten o'clock to arrive in its own plodding way, without one William T. G. Morton pushing it away with all his might.

Even as Morton was in downtown Boston hovering over Chamberlain, Dr. Warren was alone in his lab at Harvard, watching glue dry.

By surname alone John Collins Warren was an eminence in Boston in the first part of the nineteenth century. His uncle, Joseph Warren, was counted as the first military hero, chronologically, in the Revolutionary War and the first one overall in the hearts of many Bostonians. Reckoned to be the "most popular man in Boston," Joseph Warren was an established physician as the rebellion fermented in the early 1770s. Wealth, opportunity, reputation, and comfort were all laid out before him in the colony of Massachusetts, but he was willing to give it all up; what is more, he influenced others to risk their all in the Revolution, too. As it turned out, Joseph Warren was among the first to fall, dying at Bunker Hill. That made the most popular man in Boston a martyr who was beloved for generations after the war was over.

Joseph Warren's brother John (known in the family as "Jack") was another kind of war hero entirely. He did not rise to the glory of one single day but made his contribution over the course of a half dozen years by running the military hospital in Boston. As the end of the war came into view in 1781, Jack Warren was, just like his city, destitute of money but not of all hope. Asked to give a course of lectures at Harvard, he presided over the founding of the university's medical school, formally chartered in 1782.

John Collins Warren, the first of seventeen children born to Jack Warren and his wife, Abby, graduated from Harvard but never attended the medical school that might meaningfully be called his father's. He took his medical education in Europe. He then followed in his father's footsteps, proceeding to a professorship at Harvard Medical School and to the founding of an equally eminent institution, Massachusetts General Hospital, in 1811. In October 1846, however, it was Uncle Joseph Warren whom John seemed to follow, in a revolution suited to his own time.

In 1846 John Collins Warren had everything that his profession could offer, in terms of respect and, more than that, of trust. In the middle of October of that year, many people around the hospital wondered why Dr. Warren, at the age of sixty-eight, would risk that respect and that trust on an unknown entity such as William Morton and a panacea believed to be impossible: a painkiller for use in surgery. At that time the decision to stage a trial, particularly one involving a patient, lay entirely with John Collins Warren as head of surgery at Mass General. He chose to proceed.

Those who disapproved may have been right in their surprise at Dr. Warren's support of Morton—the two men shared nothing in attitude or in ethic—but they misunderstood John Collins Warren if they expected him ever to leave behind his hope for a surgical painkiller, at any age.

"What surgeon is there," he wrote, "who has not felt, while witnessing the distress of long painful operations, a sinking of the heart, to which no habit could render him insensible! What surgeon has not at these times been inspired with a wish, to find some means of lessening the sufferings he was obliged to inflict!"

Dr. John Collins Warren (1778–1856).
(Massachusetts General Hospital Archives and Special Collections)

The use of exclamation points in the above sentence is indicative of unusual emotion, for Dr. John C. Warren was not a man to use them easily. Rarely in writing and never, it seems, in speaking: The word that was most frequently used to describe him is "austere." Another was "snob."

"He was an autocrat 'enragé'* as the French say," remarked one of his closest colleagues, a physician named Henry I. Bowditch. "His own will was law for all."

Dr. Bowditch's comment was spontaneous but very, very private: While reading one of Dr. Warren's books, he'd been compelled to pick up a pencil to vent his resentment in the margin of one of the pages.

A man of many firm opinions, Dr. Warren held that most people eat too much (of course, he was probably right about that), and furthermore, that they eat the wrong things, such as fatty meats and rich desserts. In any case no one suffered the misfortune of overeating at Dr. Warren's house. He made up the menus and watched over the portions. He himself was not only thin but angular. This made him seem tall to some people, though he was not, and it certainly added to his aura of severity.

According to Dr. Daniel Slade, a former student at Harvard, Warren was a man of medium height, with a thin, somewhat stooping form. "His scanty gray hair was carefully brushed away from the high forehead," Slade said,

> the shaggy eyebrows overhanging dark sparkling eyes, while the entire expression of countenance showed determination and coolness. In manners somewhat brusque and severe, his presence was commanding, and his word was law. As a surgeon he had continued for many years to hold the first rank, a position due not only to his unimpressionable temperament, but also to his long and well-directed education.

**Enragé* did not mean "enraged" but referred to a state of mind somewhere between relentless and crazy.

As a student in Europe, John Warren had resolved never to waste a minute of the day, and it appears from his journals that he did not: He was at work from early morning until after midnight every day but Sunday, and even on that day, he would visit patients at the hospital or in his private practice. His diversions were scientific as well, and he was a member of the vaunted Boston Society of Natural History, as were doctors Gould and Jackson. Its meetings were quite formal, with presentations of papers and news, so in mid-October 1846, Dr. Warren was in the process of forming another, looser organization to allow for free-ranging discussions. The fact that he was at its epicenter was no secret. At the very first meeting the members voted to call it the "John C. Warren Club." At his insistence it was later rechristened the Thursday Evening Club.

What John C. Warren wanted to talk about most, with his fellow scientists or anyone else, was his mastodon. It may even be that the real reason for the Thursday Evening Club, as far as Dr. Warren was concerned, was to allow ample time to talk about mastodons, the prehistoric creatures that resembled large elephants. Dr. Warren happened to own one, in skeleton, which he had purchased over the summer and was preparing to display in October 1846.

While on a visit to Boston, the most famous geologist in England, Sir Charles Lyell, had seen the mastodon. "It is the most complete, and perhaps, the largest ever met with," Lyell enthused. "The bones contain a considerable proportion of their original gelatine, and are firm in texture." As Sir Charles pointed out, one of the most exciting aspects of Warren's mastodon was the condition of the bones. They did not appear to be much more decayed than any bone only recently buried. To support that observation Dr. Warren had chemical proof, supplied by his friend Dr. Charles T. Jackson.

It was Jackson's research that convinced Dr. Warren that the composition of the bones—and, moreover, their lack of decomposition—was of special significance. Warren dedicated himself to experiments with compounds that would seal the bones so that they might last as well in the open air as they had for forty thousand years in the muddy marl.

"In the morning," he noted in his journal for October 15, 1846,

> passed an hour at the Med'l College attending to the dismounting of the Mastodon. It was completed in an hour with the aid of 3 persons—the Mastodon was hardly down, when Prof. Massey of Cincinnati, recently returned from Europe, came to see it. Two important uterine cases today. Altogether the morning was confused & hurried.

The next morning, that of the sixteenth, was not hurried at all, though. Dr. Warren spent the first few hours of his day in his workroom at the Medical College, testing the effect of different types of glue on decayed bones. "White glue," he noted, "gives in some cases a beautiful appearance and the black gives great strength."

In mid-morning Dr. Warren left the Medical College for the ride through Boston's streets to the hospital, where he was expected to oversee the operation on Gilbert Abbott's neck. That was at ten o'clock.

The Massachusetts General Hospital building was located near the western slope of Beacon Hill, right along the Charles River: a mess of a waterway in those days, often left in disarray by the tides. The hospital was a sturdy edifice built of stone—that had been one of the only requisites of the Building Committee in commissioning designs for it in 1817. According to their "Resolve," the building was to be "of stone, and *of that kind* called granite [emphasis in original]." The winning design came from Charles Bulfinch, and it reflected his beautifully balanced, classical style. Rectangular in shape, the hospital building was extended into an even more pronounced oblong in 1844, with the addition of two wings to lengthen the sides. The surroundings were artfully planted with flowers and trees, and the hospital took advantage of its position on the banks of the Charles by offering facilities for swimming—when there was water in the river.

"Within the walls of the hospital the appointments were of the very best," recalled Daniel Slade, who, as a medical student at Harvard in 1846, formed a vivid impression of Mass General:

> Its wide and airy halls, its stone stairways, the scrupulously clean and well waxed floors of the wards and private rooms, its curtained beds and every article of furniture, and above all, its skilled nurses, bespoke an attention to the primary objects of the institution, and to the comfort and care of its humblest patient.

The basic rectangle of the front facade was balanced by the grand entry in the center, where heavy doors opened on either side of a pillared portico. An 1824 article described what a person would see after walking through one of those two entries, just as Dr. Warren did on the morning of the sixteenth, after his carriage dropped him off: "In the centre of the two principal stories are the rooms appropriated to the Superintendent, the apothecary, and other officers of the institution. Above these is the operating theater, lighted from the dome, and fitted up with semi-circular seats for the spectators."

The operating theater was indeed a stage, with a semicircular "proscenium." At the left was a door leading to the grand staircase, and at the right were two doors: One led to a parallel staircase, while the other, farther downstage, led to a back passageway. Along the back wall of the stage, Daniel Slade wrote, stood

> handsome cases, in which artistically arranged surgical articles of every description, and adapted to any and all description, were conspicuously placed. In one corner, in an erect posture, an Egyptian mummy, with a scarabæus on the end of its nose, looked calmly down upon the first scene of a new epoch, from its gayly adorned [*sic*], wooden sarcophagus, the cover of which had been removed.

The mummy, a gift from a Dutch trader with business connections in Boston, had been dissected by J. C. Warren in 1824, a standing-room-only event in the operating theater.

There was a skeleton in a glass case, but more ghastly by far were the hooks, rings, and pulleys set into the wall to keep the patients in place during operations.

On the morning of the sixteenth, a smattering of seats in the steeply banked gallery overlooking the operating stage were occupied by medical students. Officially the school year had yet to begin, but those students who happened to be in town were used to taking every opportunity to see an operation. The students didn't know as they sat waiting for the surgeons that there was anything special planned for that morning's operation. However, word had circulated among some of the faculty and other doctors that a surgical painkiller had been found, and at least a half dozen well-known physicians were among those in the gallery. Augustus Gould was there, in his dual capacities as Warren's colleague and Morton's landlord. Eben Frost, a patient of Morton's, was also there: he had the distinction of having had a tooth extracted under the influence of ether. Dr. Henry Bigelow, the dashing son of Dr. Jacob Bigelow, was there; both were on the staff of Mass General. Dr. Heywood, the house surgeon who had written to Morton on Dr. Warren's behalf—he was there, too.

One person in Dr. Warren's circle, and Morton's too, who was not present was Charles T. Jackson. He later said he was busy, but then that day belonged to busy men. Only one of them had as much at stake, and that was William Morton. Only one had more at stake, and that was Gilbert Abbott.

"The patient was a young man, about twenty years old," Dr. Warren later recalled of Abbott,

> having a tumour on the left side of the neck, lying parallel to, and just below the left portion of the lower jaw. This tumour, which had probably existed from his birth, seemed to be composed of tortuous, indurated veins, extending from the surface quite deeply under the

tongue. My plan was to expose these veins by dissection sufficiently to enable me to pass a ligature around them.

"The patient was arranged for the operation in a sitting posture," Warren continued, "and everything was made ready." By that he meant that the instruments to be used in the operation were arranged neatly on a table covered with a white linen cloth.

At ten o'clock, Morton was still nowhere to be seen. "Nothing had been seen or heard of Mr. Morton since my letter had been sent to him," Dr. Heywood said. In the meantime he proceeded as usual. "The necessary preparations were made, the patient fastened upon the table," Heywood recalled, "when Dr. Warren said to me, 'Stop! We have promised Mr. Morton a chance. We will give him a few minutes more."

The students, and most of the others who were there simply to observe Dr. Warren, didn't know what was happening—or why nothing was happening. "'All is ready—the stillness oppressive,'" were the words of one of those looking down on the scene from the seats, a student named Washington Ayer. Finally Dr. Warren stepped forward to address the gallery and break the stillness.

"Since many of you have not been informed for what purpose you have been assembled here," he said in a loud, clear voice, "I shall now explain it to you.

"There is a gentleman who claims he has discovered that the inhalation of a certain agent will produce insensibility to pain during surgical operations, with safety to the patient. I have always considered this an important desideratum in operative surgery," Warren concluded, "and after due consideration I decided to permit him to try the experiment."

Dr. Warren broke more than the stillness with his announcement. He broke open his own reserve, placing his own hopes on display in awe of the experiment—and his reputation within its uncertain legacy. For him the chance had been taken, but it might have been better if he had not spoken so soon.

"Those present were incredulous," reported Washington Ayer. Others reported the uncertain murmurs and quizzical looks of many in the crowd. Once the immediate reaction had receded, however, an even more uncomfortable sense of doubt was cast over everyone present, not by any fanciful hopes for a miracle but simply by the passage of time.

"A board of the most eminent surgeons in the State were gathered around the sufferer," stated Ayer, speaking of Warren and his colleagues.

The audience waited by watching that board of surgeons—which waited by watching Dr. Warren. He waited by looking at his pocket watch. "His own will was law for all," as Henry Bowditch had written. No one moved, not even the patient; no one protested the delay. Another man might have begun to feel foolish, placing so much faith in a man who had not even the grace to show up, but not Dr. Warren.

Those who spoke of his surgery as "neat, curt and *effective*," might just as well have been talking about his sarcasm . . . .

At about 10:25, John Warren turned to the gallery. The subject was obviously to be the errant dentist, William T. G. Morton: the man on whom the mighty Mass General had been prepared at that very hour to bestow the fullest use of its facilities and a profound extension of its encouragement. "It appears," Dr. Warren observed wryly, "he is otherwise engaged."

"This was followed by a derisive laugh," Washington Ayer said, "and Dr. Warren grasped his knife and was about to proceed with the operation."

Dr. Morton, still racing toward the operating dome, had but a moment left, perhaps twenty seconds at most. In terms as real as the hard stone steps leading up to the top floor of Massachusetts General Hospital, that moment was a gap. If Dr. Warren made the first cut, then he and Abbott and everyone else would be nothing more than part of the past. About twenty seconds were all there was to make that sharpest cut between one long moment and the next.

2

A BLANK WHIRLWIND OF EMOTION

In 1821 a doctor in France was told he needed an operation to remove a tumor growing in his chest cavity. His surgeon, Monsieur Richerand, proposed to remove the growth and all the parts to which it was attached: at least two ribs and part of the pleura (the membrane surrounding the lungs). It was a bold plan, but there was no choice, as the patient knew full well, since he was himself a medical man.

In fact it was an extremely bold plan. According to a report published by the Royal Society of London, "The surgeon could touch and see the heart through the pericardium [membrane sac], which was as transparent as glass, and could assure himself of the total insensibility of both." The surgeon could see and touch the beating heart.

But so could the patient, the man whose own heart it was. Without anesthetics a patient was witness to the horrifying beauty of his own life and privy to the stark secrets of his body. Even as the ailing Frenchman watched, "M. Richerand was obliged to lay the ribs bare, to saw away two, to detach them from the pleura, and to cut away all the cancerous part of that membrane." In 1821 open-heart surgery of the type performed by Monsieur Richerand was an "Extraordinary Surgical Operation," as the Royal Society titled its report—extraordinary to no one more so than the patient, the French medical man who was himself granted an extraordinary glimpse into his own anatomy—or into his own mind, depending on the object of his attention during the cutting. Possibly it was the former, because less than a month after the operation, the patient roused himself from bed just to see the fragments of his ribs, which were on display at Monsieur Richerand's office. A week after that he was back at home, engaged again in his own work.

Fully conscious or with sensibility only barely dulled by some narcotic, a patient was a partner in an operation. Surgeons took advantage of the opportunity to question patients in the midst of procedures, and there was always communication even when there were no words.

In New York City the debonair surgeon Valentine Mott had no idea what the overall effect would be when he tied a major artery to treat an aneurism (abnormal swelling) in 1818. And so he asked the patient. The operation was being performed on a sailor named Michael Bateman, aged fifty-seven.

"I drew the ligature gradually," Dr. Mott wrote, "and with my eyes fixed upon his face, I was determined to remove it instantly if any alarming symptoms had appeared. But instead of this, when he showed no change of feature or agitation of body, my gratification was of the highest kind."

"In no instance did I ever view the countenance of man with more fluctuations of hope and fear," he added.

Mott's assistant then spoke to Bateman, whose upper chest was then gaping open. He asked if the seaman felt anything unusual in those areas directly supplied by the artery, his head, breast, and right arm—"to which he replied that he did not."

Very few patients were that rational, though, and normally—in that least normal sphere—the communication was not directed at the surgeon. To whom is a scream in horror directed? John Struthers, the president of the Royal College of Surgeons in Edinburgh, described the specter of surgery without anesthesia by speaking of "the operating theater ringing with the groans and shrieks of the patient, the distressed faces of the crowd of students, and the haste of the operator to be done." By the first half of the nineteenth century, surgery remained as a relic of ancient agony stranded by itself in a civilization that was modern or quickly modernizing in every other respect.

"The amputation of a limb is an operation terrible to bear, horrible to see," stated Percival Pott, a surgeon at St. Bartholomew's Hospital in London in the mid–eighteenth century. Nonetheless, amputation was among the most common of surgical operations in the first part of the nineteenth century, second only to removal of surface tumors, at major hospitals. It was almost certainly first in frequency overall, taking into account the work of country doctors, such as one who told his new assistant of a farmhouse amputation he performed in about 1805: "I sharpened a cheese knife and borrowed the carpenter's saw, and got through it pretty well."

The ancients might well have jeered. Up-to-date surgeons in those early 1800s offered patients very little for the passage of years: thousands of years that had made new realms of geography, engineering, transportation, finance, and chemistry, industry, and even general medicine. In every field practical sciences were rising to the challenge of overcoming those tribulations that might once have been classified as the "human lot." By the nineteenth century, those that were still left seemed permanent to that human lot, even constituting, perhaps, the essence of it. There was war; no advance had changed it, at its base. There was poverty; nothing dislodged it, least of all prosperity. And there was pain, which was wrapped tightly in "God's will," along with two other unarguable events of the human mystery: death and childbirth.

People in the early 1800s were living longer and better, but they were dying just as before if they were facing a surgeon's knife. Surgery was forced to remain a social concern, not a scientific one, because what mattered in preanesthetic days was, first, deferring to the agonies of the mind, and only second, tending the ills of the body. Surgical science was a critical part of humanity's progress, but it was bound fast to an even deeper part of the human lot—the inescapable fact of pain.

On the day of an operation, a patient was led up to the top floor in any hospital with a dome or a tower—or down to the basement, if there was no tower. Towers offered the best light, but that is not why they housed the operating rooms in modern hospitals of the early 1800s: They isolated the sound of screaming. Tables were installed to accommodate patients, but regular chairs were used just as often, since the patient was conscious and could therefore remain upright. One of the sights that might strike terror into the patient even before the operation began was the operating costume adopted by the surgeon. The custom was to wear an overcoat, usually an old winter coat, and to leave it unwashed between operations. Naturally it became stiff with the dried remains of all the fluids that spurt and gush during surgery—mostly blood. In some quarters it was perceived as a badge of experience to have the coat with the thickest caking of old matter.

The equipment in the modern operating room of the early nineteenth century was neat, if not antiseptic. High-ranking hospitals in France, as well as Italy, Germany, England and Scotland, and the United States, boasted clean and airy operating theaters, where fresh white linen cloaked the tables (if not the surgeons). Gleaming instruments would be laid out in order before each operation, much as today, with one difference: The gleam came from the handles, which were polished between operations. The blades were left dirty.

Addressing the issue of pain, surgeons commonly prescribed opium to reduce the patients' sensitivity during surgery. That was a long-standing practice, started in ancient times. Patients, solely at their own discretion, could take some number of drops an hour or so before an operation. The results were not auspicious, though, ranging from mere nausea—which bore no effect on sensitivity—to death, which bore rather a complete one. More often people took liquor before an operation. Surgeons pointed to examples that proved how effective true intoxication could be. Dr. Francis Boott, a well-known American doctor living in London, wrote of a colleague who would frequently refer to "the case of an Irishman, part of whose face was eaten by a pig while he was lying dead drunk on the ground." Alcohol, however, offered imperfect pain relief in surgery for the majority of patients.

Whatever the means used during an operation, the goal was to put the mind in a state in which it could not feel anything. Some people tried to jolt themselves chemically into such a manner of thought, in which all information from the senses would somehow be dissipated as soon as it was delivered to the mind. They were the ones who looked to liquor, opium, or any number of other herbal drugs, none of which worked reliably or completely.

Some people went into surgery certain that they could impose a formula of thought upon themselves: that they simply would not absorb any report of pain. They were the stoical ones, congratulated by their relatives and heartily welcomed by the surgeons.

A third means of erasing the sensation of pain was mesmerism, better known as hypnotism. It was invented by a wealthy Swiss named Franz Mesmer, who used magnets, incense and soft words to lead patients into an insensible state—a "trance," in the parlance of the genre. Most physicians thought it was utter bunk, but a few would entertain it. It flopped in the majority of cases; surgeons couldn't object to that—so did everything else that was supposed to render patients insensible. Surgeons disdained mesmerism because it took up to twenty-four hours of tedious mesmerizing to produce a trance. That was awfully inconvenient. And, of course, mesmerism didn't smack much of science: Nothing involving incense, ever does.

Because the mind was too complex to understand, let alone to train, some surgeons kept their distance from it and developed a form of local anesthesia. The most common was choking off the flow of blood to the area which was to be cut, rendering it numb. Another technique called for the application of ice, which resulted in approximately the same type of numbness. It was said to have been originated by the French military surgeon Jean Dominique Larrey, who noticed while on campaign in Russia that soldiers whose limbs were not only wounded but frostbitten complained very little of the pain of cutting.

James Wardrop, an Englishman who was no less than the Prince Regent's personal surgeon in 1821, congratulated himself on his response to one young woman "of robust form," suffering with a tumor on her face. She became so frenzied at the first touch of the scalpel during her operation that the effort had to be abandoned. "As the only recourse," he wrote, "it then occurred to me, that if she would allow herself to be bled to a state of deliquium,* the tumor might be extirpated while she remained insensible." As fifty ounces of blood drained from her arm, the patient fainted, and the operation proceeded at last. What is more, Dr. Wardrop boasted that her recovery was greatly enhanced by the enormous amount of blood she lost. His technique caught on in the 1820s and 1830s and was known as "bleeding to *deliquium animi.*"†

In the same era the most respected surgeon in France, Alfred Velpeau, was advising his students that "the surgeon must do everything, albeit without overstepping the limits of truth, to induce the patient to ask for the operation himself." Velpeau maintained that it was a surgeon's "duty" to trick patients "as to the length and the severity of the pains which they would experience, as well as the dangers they would be exposed to."

Many surgeons who were above the ruse of actually fooling patients into accepting surgery did fall in with the professional habit of reporting patient reactions only when they were duly tranquil. An English surgeon published an article about his treatment of a hunting wound sustained by one of his patients, aged fifty:

---

*Fainting.

†Suspension of the life force.

> At his request, on 23rd September, 1833, I removed the whole of his left eye-ball, with its lachrymal gland, and divided the optic nerve far back in its socket. . . . Such was the patient's extreme fortitude and perseverance that not even was his hand raised, nor a syllable of complaint uttered during this most painful operation.

That was admirable, but it wasn't usual.

A doctor at Massachusetts General told of an experience in which he was the one with the "fortitude and perseverance," and also the one with his hand raised. His patient came in quaking, took one look at the operating room, and then bolted down the hall, locking himself in the bathroom, screaming the whole time. The surgeon gave chase. He apparently didn't think there was much chance of inducing the man to ask for the operation himself (as Velpeau had suggested), and so he broke down the door and captured him by the neck. He then dragged the patient under his arm to the operating theater, where the surgery proceeded, without further rugby.

"I do not wonder that the patient sometimes dies," said Edward Everett, a statesman and orator, "but that the surgeon ever lives."

Dr. Wardrop advised that in preparation for an operation on an extremity, children and other recalcitrants be enclosed in a bag or a box, with the part to be cut left out through a hole. The patients could react any way they wanted, as far as Wardrop was concerned, as long as they couldn't kick him in the stomach in the process.

Whimpering, praying, mumbling, moaning, cursing, wailing, shouting: Those were the understandable lot, but some number of patients facing surgery without anesthesia slipped to the side with a quiet worse than all the screams in a hospital dome. It is impossible even to say what that number was exactly, except that in the experiences related by every surgeon, there were those who refused surgery. They chose death instead, either by suicide or acquiescence with the course of disease.

An aging bachelor, recalled in an article in the *Dublin Review* in 1850, was said to have borne heartbreaks and injustices in his life gracefully—and more than that, with "stoical pride." When an operation became necessary, though, "He went home from the study of the medical man who had told him so, made a kind and rational will, and the next morning destroyed himself."

John Collins Warren also told of a patient in Boston who committed suicide in horror at the prospect of an operation; most surgeons knew of such cases. The victims were driven to some sort of insanity, seeing their surgery even more vividly than they saw their own death. Others, however, survived in their minds by refusing to look at either scene.

Among those many who took the second course was Georges-Louis Leclerc, comte de Buffon, the French naturalist who cast himself as a leader in the new belief in science in the eighteenth century. To him zoology reflected the beauty of science—of cruel reality, not miracles. Nonetheless, at the peak of his fame, Buffon refused an operation for bladder stones. Help was available, but at a cost he could not accept, and a belief in science over self that even he did not quite have. He died, still in need of surgery, in 1788.

The operation for bladder stones, called a lithotomy, involved probing through delicate tissue along the urinary tract. Because the difference between an expert lithotomy and an average one was measured in units of sheer agony, everyone—without a sane exception—wanted the very finest surgeon. Stories about the others emerged occasionally to underscore the point. "Surgeons," wrote a Philadelphia physician named John Syng Dorsey, "have been baffled in their attempts to find a stone with the forceps . . . and sometimes an hour has been consumed in fruitless searches for the calculus [mineral deposit]." A good lithotomist could complete the operation in about one minute.

Each city had its lithotomy specialist—there was no second best. John Collins Warren claimed without boasting that he performed every lithotomy rendered in Boston from 1804 to 1844. In New York City the lithotomist during the same years was

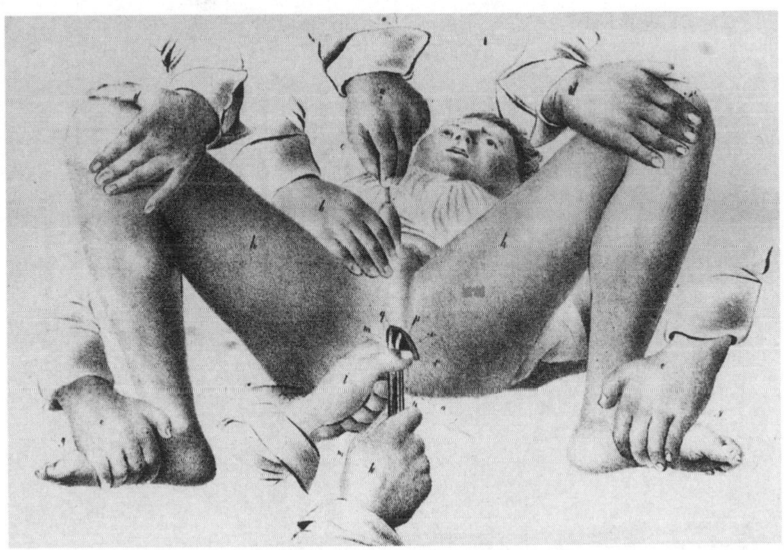

A lithotomy circa 1837—before anesthetics.
(Library of the College of Physicians of Philadelphia)

Valentine Mott. In Philadelphia, which traditionally served the most critical medical needs of the entire South, an eminent doctor with a well-suited name, Dr. Philip Physick, was the lithotomy man for years. It was Dr. Physick who was chosen to perform the operation in 1831 on John Marshall, Chief Justice of the Supreme Court and the government's most influential figure. John Syng Dorsey, Dr. Physick's nephew, eventually inherited the Philadelphia mantle.

Dr. Dorsey listed a number of tricks toward locating the stone or stones as quickly as possible, and included a reminder as to why no one wanted a beginner as a lithotomist. "The surgeon should always remember, that it is better to *cut* than to *tear*," he wrote, underscoring the point with an old bit of wisdom from Celsus, a first-century surgeon who wrote the earliest known essay about the lithotomy. Dr. Dorsey advised that the maxim of Celsus should never be forgotten: *Plaga, paulo major quam calculus sit* (The wound should be a little larger than the stone).

However people may have acted outwardly, their senses were

Dr. Philip Physick (1768–1837). (Pennsylvania Hospital)

bound to be stretched to snapping during surgery. An operation without anesthesia was nothing more than trauma at the top of the hour: on a schedule. The primary sensation was that of feeling, but those of sight and sound were even more haunting. When one speaks of "pain" during an operation without anesthetics, it is a word with ragged tails of meaning and imagery that permanently dye the mind: the peculiar red of one's own blood, the echoing blue of a limb dropping to the floor.

A correspondent who wrote a lengthy letter to the mid-nineteenth-century Scottish surgeon James Simpson left what is generally considered the most vivid and detailed account of surgery without anesthesia, from the patient's point of view. He was never stoic but ever human. "The operation was a more tedious one than some which involve much greater mutilation," he wrote. "It necessitated cruel cutting through inflamed and morbidly sensitive parts, and could not be despatched by a few strokes of the knife."

> Of the agony it occasioned I will say nothing. Suffering so great as I underwent cannot be expressed in words, and fortunately cannot be recalled. The particular pangs are now forgotten, but the blank whirlwind of emotion, the horror of great darkness, and the sense of desertion by God and man, bordering close upon despair, which swept through my mind and overwhelmed my heart, I can never forget, however gladly I would do so. Only the wish to save others some of my sufferings makes me deliberately recall and confess the anguish and humiliation of such a personal experience; nor can I find language more sober or familiar than that I have used to express feelings which, happily for us all, are too rare as matters of general experience to have been shaped into household words.
>
> During the operation, in spite of the pain it occasioned, my senses were preternaturally acute, as I have been told they generally are in patients under such circumstances. I watched all that the surgeon did with a fascinated intensity. I still recall with unwelcome vividness the spreading out of the instruments, the twisting of the tourniquet, the first incision, the fingering of the sawed bone, the sponge pressed on the flap, the tying of the blood vessels, the stitching of the skin, and the bloody dismembered limb lying on the floor. Those are not pleasant remembrances. For a long time they haunted me, and even now they are easily resuscitated; and though they cannot bring back the suffering attending the events which gave them a place in my memory, they can occasion a suffering of their own, and be the cause of a disquiet which favors neither mental nor bodily health.

Expertise and futility, ambition and stagnation: Perhaps the image most symbolic of the art of surgery in the years before the introduction of anesthesia is that of dedicated surgeons, practicing complex, lifesaving operations—on cadavers. Living patients couldn't withstand prolonged or intricate cutting. Surgery was considered as a last resort, after all else in medicine had failed. The working theory in the first decades of the nineteenth century was that the true triumph of medicine would result in the obsolescence of surgery. The only wonder was that such excellent men were drawn to it at all.

Robert Liston was giving a lecture on operative surgery to the students at Edinburgh's University College Hospital in 1844, when he observed that it "is regarded as an inferior part of our professional duties.

"And truly it is so." he reflected. "The field of operative surgery, though happily narrowed, is still extensive." (Fifty years later, J. Collins Warren, namesake of his grandfather John Collins Warren, retold that story with fresh astonishment: "What a contrast!" he exclaimed in a lecture of his own. "A great surgeon *rejoicing* that the field of operative surgery was *narrowed*!")

Those within the world of surgery had no choice but to accept the aura of dismay constantly around them.

The esteemed English surgeon Astley Cooper once related an anecdote about his own uncle, William Cooper. "He was going to amputate a man's leg," Sir Astley wrote of his uncle William,

> when the poor fellow, terrified at the array of instruments and appliances, suddenly jumped off the table and bolted off; seeing which, the operator, instead of following the man, and attempting to persuade him to submit to the evil which circumstances rendered necessary, turned round and said, apparently much relieved by his departure, "By God! I'm glad he's gone."

"My uncle," added Sir Astley, "was a man of great feeling—too much to be a surgeon."

3

THE HILARITY BEFORE ETHER DAY

In 1833 a newspaper in Albany, New York, printed a rave review of a demonstration of laughing gas (nitrous oxide) presented by an itinerant lecturer named Dr. Coult—Dr. Coult of London, New York, and Calcutta.

As a matter of fact, though, Dr. Coult was not quite a doctor. He was Samuel Colt, a tall nineteen-year-old from Connecticut who had already attended a boarding school in Amherst, Massachusetts, dropped out of the merchant marine, quit his father's small textile factory—twice—and, more important, completed the designs for the first practical revolver-type pistol. He had indeed been to New York, London, and Calcutta (as a seaman), but he hadn't had to go nearly that far from home to learn about nitrous oxide. Whenever he and the chief chemist at the textile factory grew bored, they did what many laboratory workers of the day did: they indulged in a whiff of laughing gas.

"The effect which the gas produces upon the system," reported the *Albany Microscope*, "is truly astonishing. The person who inhales it becomes completely insensible and remains in that state for about the space of three minutes, when his senses become restored.

"We never beheld such an anxiety as there has been during the past week to witness the astonishing effects of Dr. Coult's gas," the *Microscope* added.

In order to raise the money with which to patent his revolver, Sam Colt had taken to the road in New England with nothing more than a cart loaded with the apparatus needed for the manufacture of nitrous oxide. It was the lowest start imaginable: Pausing on a corner in any likely village, he would demonstrate the gas on himself and then sell snorts for a quarter. Eventually he made his way west to Cincinnati, where his brother John was partaking fully in the life of the free-spirited city. Through John Colt's popularity and Dr. Coult's promotional efforts, laughing gas became the centerpiece of a permanent exhibition in its own small theater in Cincinnati, with two or three shows daily. As a matter of fact, John's residence was itself something of a permanent exhibition, a party that rolled along for people of all sorts, looking for something new—something just like laughing gas. Sam became a convivial figure, even without much wanting to be. Examining the situation, he decided that for the sake of his future he'd better depart Cincinnati.

For Colt the gas represented nothing more than the quickest possible way to finance his invention, the revolver. To him, as to most of science in the 1830s, nitrous oxide was nothing more than a trained bear, to be taken around and shown for money. But, like the bear, it was not understood.

By the 1830s laughing gas certainly didn't belong to the world of medicine. The closest it may have come was the occasion when Dr. Coult was working the Mississippi and happened to be a passenger on a riverboat in the throes of a cholera panic. The passengers praised God: There was a doctor on their ship. When they besieged Colt's cabin, though, he had a panic of his own. Eventually, and very reluctantly, he saw the sick and doled out the only palliative he knew, which was laughing gas. The patients all recovered nicely, leading them to spread word of a medical miracle, and leading Sam Colt to conclude that none of them had had cholera in the first place.

When Colt retired from laughing-gas entertainment at the age of nineteen, he had the two vital components of destiny, at least for a Connecticut Yankee in 1835: narrow-eyed tenacity and a U.S. patent. Over the following two decades his Colt revolver would become one of the soaring successes of American manufacturing.

A generation before Colt's, in the very first years of the nineteenth century, nitrous oxide was not yet the stuff of theatrical entertainments. At that time, on the contrary, it was the subject of more serious studies than any other compound. Chemists devoted important books to their experiments with nitrous oxide. Poets followed close behind the chemists—and satirists followed the poets. And opportunists like Sam Colt followed in a crowd after that. It was a wayward path, one that kept surgery in a state of disarray, unnecessarily, for almost fifty years.

Nitrous oxide had been discovered in 1772 by the versatile Joseph Priestley, an English historian and theologian who also isolated oxygen in the course of his short sojourn into the field of chemistry. Priestley's work with gases was relegated to chemistry books until 1799, when another Englishman, Thomas Beddoes, launched an effort to use the new gases in medicine. Beddoes was a prominent physician in Bristol, a port city in the west of England. Though not a brilliant man, he was no doubt an energetic and savvy one, whose great goal was to apply "the different kinds of air" discerned by Priestley toward the relief of respiratory diseases. In Beddoes's time the British population was ravaged by tuberculosis and a whole array of other diseases that affected breathing. In many European countries respiratory illness had become the leading cause of death among adults, a situation that developed in tandem with polluted and crowded conditions. Those conditions may have nurtured the crisis, but the lack of effective remedies also contributed to the death rate, and that was the angle from which Beddoes attacked the problem.

"Never be idle, boys. Let energy be apparent in all you do. If you play, play heartily, and at your book, be determined to excel. Languor is the bane of intellect." That was the advice that Dr. Beddoes gave to two schoolboys named Lambton who had been placed in his care by their wealthy father. It might also be a kind of self-portrait of Thomas Beddoes.

In 1798 Dr. Beddoes decided that languor was also the bane of public health, and he conceived a new type of institution, a private laboratory where scientists could dedicate themselves to research, while expenses were underwritten by humanitarians—very rich ones. The new laboratory-clinic pioneered a partnership between science and private philanthropy. No sooner did Dr. Beddoes happen to mention the plan to his friend William Lambton, than he had in hand the gigantic sum of fifteen hundred pounds with which to launch the effort. Josiah Wedgwood, the china manufacturer, also made donations in support of the new clinic, which was to be called the Pneumatic Institution.

During the planning of the Pneumatic Institution, Dr. Beddoes received a letter recommending a penniless chemist then working in Penzance, Cornwall, a man by the name of Humphry Davy. Though only twenty-one, Davy was engaged to be the superintendent of the new institution.

Rakish in appearance, lean and dark, Humphry Davy became a very popular figure around Bristol, and was soon known throughout England. The institution he headed captured the imagination of the populace, which was easily persuaded that an outright end to respiratory disease was not too much to expect, with the country's best minds at work on the problem.

As superintendent, Humphry Davy chose nitrous oxide as the first gas for investigation. No sooner had that decision been made than he announced his intention to inhale it himself, despite the fact that it had been labeled poisonous by all previous researchers, including Priestley. That did not concern Davy at all. "He seemed to act as if, in case of sacrificing one life, he had two or three others in reserve on which he could fall back in case of necessity," observed his friend, the publisher Joseph Cottle.

In 1800 an American journal, the *Medical Repository,* introduced the Pneumatic Institution to readers in the United States, calling it "as useful as novel." The journal further submitted that it would hardly be possible to find anyone better qualified as superintendent than Humphry Davy.

However, at any hour of the day or night during late 1799 and early 1800, one would be likely to find that young genius, the superintendent, running around the Pneumatic Institution giggling ferociously. A moment later he might be lying in some hallway, sound asleep. Just as likely he would be surrounded by friends, fashionably dressed and temporarily insane.

After proving by experiments on himself that nitrous oxide was not only safe but rather sublime, Davy often inhaled the gas with his friends. The effect depended, then as now, first on the amount taken and second on the disposition of the person while inhaling. Those who were excited usually had that feeling exaggerated. Humphry Davy was unique in taking the gas when he was in a relaxed state, inhaling enough to induce sleep; in fact he claimed that taking the gas that way cured his recurrent headaches. Others took it in smaller amounts and with the expectation of the peculiar inebriation. The effect was not pleasant for everyone, though. Neither Josiah Wedgwood nor his brother Thomas enjoyed it particularly, but Samuel Taylor Coleridge did—immensely. So did another poet, Robert Southey, who made a remark often quoted in nitrous oxide lectures. "The atmosphere of the highest of all possible heaven is composed of this gas," he claimed.

One of Davy's patients at the institution was even more lyrical—and certainly more spontaneous. Asked about the effect of the gas, the patient replied, "I felt like the sound of a harp."

In 1800 Davy collected everything he knew about nitrous oxide into a widely read book called *Researches, Chemical and Philosophical; chiefly concerning nitrous oxide.* It was no mere pamphlet but stretched to more than four hundred pages. While he could not name any definite use for the gas, he did cite his own experiences with insensibility under its influence when he made his suggestion, the one so often quoted in retrospect, that nitrous oxide "may probably be used with advantage during surgical operations."

Sir Humphry Davy demonstrates nitrous oxide for his friends.
(Thomas Gilray drawing courtesy of Wood Library-Museum, Park Ridge, Illinois)

There is no record of anyone acting on Davy's idea in a surgical operation. When hope is long gone, as it was in the case of eliminating pain during operations, even the plain and obvious demands too much imagination. After Davy's suggestion had been out in the open for a full generation, Alfred Velpeau was still assuring his students in Paris in 1839 that:

> To escape pain in surgical operations is a chimera which we are not permitted to look for in our day. Knife and pain, in surgery, are two words which never present themselves the one without the other in the minds of patients, and it is necessary for us surgeons to admit their association.

That is what surgeons did, even though many of them were watched all the while by a copy of *Researches, Chemical and*

Philosophical, chiefly concerning nitrous oxide, standing on their shelves.

While Davy and other scientists looked for something—anything—that nitrous oxide would actually *cure,* socialites and the literary types who blur into the social world found that it was a perfect remedy for boredom. At the height of fashion at fashion's height—London, England, in the Regency era—the gas was soon offered at dinner parties instead of wine. Serious poems and popular songs were written about it. Being a fad, nitrous oxide naturally caught the interest of the humorists as well.

William P. C. Barton, a serious-minded chemist studying in Philadelphia, deplored the way that the reputation of nitrous oxide gas sank so quickly into derision. Referring to the skepticism attached to the new gas, he wrote in the preface of a book-length dissertation on nitrous oxide in 1808, "Experiments repeatedly performed by Mr. Davy, and confirmed by other chymists, did not remove the distrust. Poetry indeed was enlisted to expose the delusion, as it was termed, and to laugh it into contempt."

A little contempt, as it turns out, sold more books than did dissertations. Thomas Green Fessenden, a native of New Hampshire, was living in London in about 1800, and he wrote about the excitement that surrounded nitrous oxide gas, especially in pseudoscientific circles. The result was a bestselling book of clever verse, profuse with footnotes that were even cleverer, to ensure that the reader savored every nuance within the verse. For the book, called *Terrible Tractoration,* Fessenden took on the persona of "a visionary, eccentric, would-be philosopher, endeavoring to effect 'grand discoveries and inventions.'" In other words, a man very much of his day.

[His plan to write a poem is dashed by the realization that he can't write poetry. . . .]

What then occurs? A lucky hit—
I've found a substitute for wit

. . . .

Beddoes (bless the good doctor) has
Sent me a bag full of his gas
Which snuffed the nose up, makes wit brighter,
And eke* a dunce an airy writer.

[He inhales nitrous oxide . . .]
How swiftly turns this giddy world round,
Like tortur'd top, by truant twirl'd round;
. . . .
I'm larger grown from head to tail
Than mammoth, elephant, or whale!—
Now feel a "tangible extension"
Of semi-infinite dimension!—

Inflated with supreme intensity,
I fill three quarters of immensity!
. . . .
But now, alas! a wicked wag
Has pull'd away the gaseous bag:
From heaven, where thron'd, like Jove I sat,
I'm fall'n! fall'n! fall'n! down, flat! flat! flat!

After two centuries Fessenden's *Terrible Tractoration* still retains its satiric sting, but the damage done by it and many other similarly insouciant works can be seen as having an even greater sting in the delay of the use of nitrous oxide in medicine. For almost fifty years the gas was nothing more than chemistry's plaything: a good laugh, the subject of dissertations, the headliner on playbills, and ever a good four blocks from any operating theater. Seen in retrospect, that gap of nearly fifty years changed the very experience of surgery; what had seemed like pain for patients was throughout that span violence—a kind of violence perpetrated by history.

---

*Makes larger.

When anesthesia was finally heralded, it was referred to as "one of the most eminent of the benefits yet bestowed upon suffering humanity." But humanity suffered in 1800 too. It yearned for miracles then. It always would. In 1800, though, it plainly wasn't ready for anesthetics, the easiest miracle of them all, set out in print by Humphry Davy. Instead the strange gas disappeared into the laughter.

As nitrous oxide moved through the social settings of the early 1800s, it was considered harmless, although it did come with certain caveats.* If improperly made the gas could easily retain impurities that would cause unintended reactions. In any case common air had to be inhaled in proper amounts, along with the gas. Nitrous oxide is not physically addictive, though sporadic reports of psychological dependence have surfaced through the years. But it was not illegal, and elders in the first two decades of the nineteenth century considered it to be acceptable—a homemade circus for teens.

For their part fashionable young people partook in a nitrous oxide party whenever they could. One account of the use of the gas in polite circles was included in a letter written by a young teenager named Anne Warren in 1825. Miss Warren, a relative of the surgeon John C. Warren, was enrolled as a student at the school for young ladies operated in Boston by Ralph Waldo Emerson and his brother. Her upbringing was careful and controlled, and she was comfortable within it. "I drank tea at Aunt Warren's yesterday," she wrote, referring to John Warren's wife. "Cousin Abby and Edward made some gass [*sic*] in the afternoon expecting two gentlemen in the evening to take it, but they did not either of them come and Cousin Edward had to take it all alone. It had but very little effect upon him. It only made him laugh and walk about the room."

For part-time thrill seekers, the only problem with nitrous

---

*Today it is known that prolonged exposure can cause cancer.

oxide was that the process of making it was cumbersome, requiring laboratory apparatus and expertise enough to reduce the chance of impurities. Under the circumstances the majority of laughing gas parties were held among medical students.

The class at the Fairfield Medical College in New York State was looking forward to a nitrous oxide party that was to be held November 5, 1838: "A goodly number of students intend to take it," one student wrote the day before, "and we expect some sport as usual." The gas was a traditional release for students cloistered for long months in dissecting rooms and lecture halls. The professors rarely paid attention to such antics.

In 1822, though, Benjamin Silliman took note of two unusual cases involving the use of nitrous oxide at Yale. Both were surprising and in different ways, disturbing. "For several years," the famous professor of chemistry wrote in the *American Journal of Science and the Arts,* "the medical class, and the two senior academical classes in Yale College, while attending the chemical lectures, have been in the habit (each class by itself) of preparing for themselves, and administering to their respective members, the nitrous oxide, or exhilarating gas." In one case, Silliman reported, a cheerful nineteen-year-old with the initials A. B. became delirious and violent after trying the gas. "For the space of two hours," ran the report, "he was perfectly unconscious of what he was *doing*, and was in every respect like a maniac; he states however, that *his feelings* vibrated between perfect happiness, and the most consummate misery." While his classmates worried about him, A. B. spent the following four days prostrate, in a state of nervous exhaustion. According to Silliman, A. B.'s case stood as a caution against exhilarating gas, "especially to those of a sanguine temperament."

Some people, it seems, don't need to be exhilarated. The second Yale case, though, was practically an endorsement of nitrous oxide for those who do.

C. D. were the initials of another student who tried nitrous oxide with his classmates in 1822. According to Silliman's description, "For nearly two years previous to his taking the gas, his health had been very delicate, and his mind frequently gloomy and depressed." After inhaling three quarts of the gas, C. D. was energetic and humorous to a degree unusual in anyone, but "astonishing" in him. Moreover, the use of the gas gave him a sweet tooth, a fact that fascinated Silliman: "Although nearly eight weeks have elapsed since he inspired the gas, he is still found *pouring molasses over beef, pork, poultry, potatoes, cabbage, or whatever animal or vegetable food is placed before him.*" (The emphasis is Silliman's.) Rather less astonishing, to the professor at least, than the fact that C. D. poured molasses over everything was the fact that three quarts of nitrous oxide had permanently restored C. D.'s health, strength, and ability to concentrate and had left him "habitually cheerful, while before, he was habitually grave, and even, to a degree, gloomy."

Both C. D.'s transformation and A. B.'s scare were anomalies, caused perhaps by their predisposition or by impurities in the gas; it is impossible to say, of course. At the time the two Yale cases only showed even more distinctly than others nitrous oxide's curious power—to let go the mind.

Young hedons without access to a lab, but with some extra cash, could hire a pharmacist or chemist to supply laughing gas for a party, bringing it in bags all ready for the guests. That didn't happen often, though—because an ether frolic was so much less bother.

"College boys and factory girls had inhaled Ether with the utmost freedom, without any ill effects upon their health," admitted Dr. Charles Jackson, recalling the 1820s and 1830s.

Through the first decades of the 1800s, one experimenter after another posted notices claiming to have discerned a remarkable similarity between the effects of nitrous oxide gas and the vapors of sulfuric ether. To a chemist long past the age of fun, that might have counted as a discovery. To any self-respecting teenager in town, though, it was old news.

Nitrous oxide and ether: party makers for the innocent in the first half of the nineteenth century. The events leading toward Ether Day would ultimately stretch out like a long length of twine because of the similarity in the effect of nitrous oxide and ether—making progress by crossing from one to the other.

Sulfuric ether, which is a clear liquid, had been known since the mid–fourteenth century but had entered medical use only in the 1760s. "Now to the medicines in Vogue . . . ," wrote an Englishman named George Cuthbert to a fellow doctor in New England, on February 16, 1760. First he discussed the latest cure for venereal disease. It was an extract of sarsaparilla, which "was supposed to be so efficacious a Remedy that there was no more difficulty in curing a confirmed pox with all its dreadful symptoms than in healing a cut finger." It didn't work, however, as Cuthbert noted. "The next to this is a medicine called aether [*sic*]," he continued. "They go so far as to give it internally for all pains and complaints, but I fear it do's [*sic*] more harm than good."

Eventually the pharmacology of ether was narrowed, and it was prescribed only in cases of breathing difficulty or nausea. It was also said that a few whiffs of ether could sober up a drunk. Its fumes could be inhaled, usually from a handkerchief, or it could be taken as a drink.

Dr. Thomas D. Mitchell, a native of Pennsylvania who taught medicine at the College of Ohio, paused to reminisce about ether in the 1832 edition of the textbook he wrote, *Elements of Chemical Philosophy*. "Some years ago," Dr. Mitchell wrote,

> a practice obtained among the lads of Philadelphia, of inhaling the vapor of sulphuric ether, by way of sport. A small quantity, placed in a bladder, was almost instantly converted into vapor, by the application of hot water. By means of a tube and stop-cock, the gas could be easily inhaled. In some cases, the experiment excited mere playfulness and sprightly movements; but in several cases, delirium and even phrenitis [brain fever] was induced, which terminated fatally.

Whatever the proportion of youngsters who heeded Mitchell's dire warning and abstained from ether use, the rest no doubt benefited from his recipe.

"Ether is an active stimulant and anti-spasmodic," explained Professor N. Chapman of the University of Pennsylvania in his 1825 textbook, repeating what medical students of the day would have heard on the subject of ether. He said it was

> somewhat analogous to alcohol in its leading effects, though more powerful and less permanent. It is sometimes prescribed in low states of disease and particularly in typhous[*sic*] fever. But its impressions are so evanescent that little is gained by it, and it is difficult to imagine a case in which it should supersede wine, etc.

"Lately an opinion has been advanced," wrote Prof. Chapman, preparing to quote another article and edging as close as he ever would to a truly great discovery, "'that Ether acts *directly sedative* on the spinal system,' the foundation of which seems to rest chiefly on the conspicuous relief afforded in a case of tetanus [which causes muscular rigidity]."

Chapman's readers were evidently much more intrigued by ether's inebriating effect than by the fact that it could be a sedative. Ether frolics lit up the social scene for teenagers all over the country. Where laughing-gas parties were necessarily held in labs most of the time, ether frolics blossomed right in the front parlor. Young men and young ladies attended together, and as though that fact was not exhilarating enough for most of them, ether fanned the air. Dr. J. Miller of Louisville, Kentucky, reported that in his region in the 1840s, ether was being used frequently among boys at school and young people of both genders at parties.

Whenever it is that self-abandonment is no longer natural but must be induced, childhood ends, and from the 1820s to the 1840s, ether frolics were regular events in many communities around the country, especially for those old enough for self-abandonment but not quite old enough for whiskey. Yet there was still something dangerous about the innocent ether frolic, because sulfuric ether had its own hidden capacities for harm. In the first place it made some people throw up—a grave tragedy at any party. Moreover, it made some people unconscious, "dead to the world," in an expression that originated in the day.

In 1818 Benjamin Silliman cautioned that experimentation with sulfuric ether was potentially very dangerous. The case that he cited in support fell close to the state of surgical anesthesia: "By the imprudent inspiration of Ether," Silliman wrote, "a gentleman was thrown into a very lethargic state, which continued with occasional periods of intermission for more than thirty hours." That case was originally cited by J. Pereira, the Spaniard who organized the pharmacology of the early nineteenth century in his book, *Materia Medica.* Pereira's conclusion was: "If the air be too strongly impregnated with Ether, stupefaction ensues."

Pereira and Silliman, Mitchell and Chapman, and many thousands of people who read their works *almost* knew. They nearly knew but bowed low before a gap that kept them from realizing that ether's stupefaction was surgical anesthesia.

Medicine couldn't possibly advance in an era that bound it so perfectly in ignorance. That is not "ignorance," as in lack of knowledge, but literally "ignore-ance"—ignoring the knowledge that was there. Medical books made glancing references to ether's sedative effects, to the lethargy and stupefaction it caused, while up in an operating theater in a Scottish hospital, an old woman was put through a mastectomy, coldly conscious of every slicing cut. She made such a pathetic sight that one student who witnessed the operation was impelled to quit medical school on the spot.

Then, in 1844, for no particular reason, a happy-go-lucky man from Vermont, Gardner Colton, decided to go on

Gardner Colton (1814–1898). (Wood Library-Museum, Park Ridge, Illinois)

Broadway, taking to the stage there. Medicine should praise the day that he did. He put nitrous oxide in a vehicle that was headed somewhere.

At once cheerful and erudite, Colton was brand new to show business when he staged his "Grand Exhibition" in New York City on March 19, 1844. For his part he might object on that point and say that he never was in show business, that he was a lecturer in chemistry—but it makes little difference, because Gardner Colton was also new to chemistry lecturing when he staged his first Grand Exhibition. At twenty-nine he was a medical student on the verge of quitting for lack of money. Colton once speculated about how some future biographer might describe his early life, which started with an apprenticeship as a chair maker: "He next studied medicine," Colton mused, "but as he never practiced the profession, he didn't kill anybody."

Gardner Colton was born in Vermont in 1814, the youngest of twelve in an impoverished family living in the country. His parents didn't do much except work and read the Bible—and perhaps more of the latter than of the former. "It was thought by some," Colton later explained, using carefully chosen words, "that in providing for his family and the future, [F]ather placed too much reliance on Providence and too little on his own exertions." As a second-year student struggling to meet the tuition at New York's College of Physicians and Surgeons, Colton earned some extra money by giving talks on chemistry at a young ladies' seminary in the city. One of his talks included a demonstration of laughing gas, and when his fellow medical students heard about the frivolity that ensued, they inveigled him into repeating the experiment for them, which he did.

A friend who was carried away by the fun insisted that Colton repeat the frivolity but on a much bigger scale—that he "bring out a grand exhibition in the great Broadway Tabernacle." Though it was just something merry to say at a party—Gardner Colton didn't have money enough to continue medical school, let alone start staging grand exhibitions in the biggest theater on Broadway—"the thing got into my head," he confided later.

"I determined to carry it out," he went on. "I went to Mr. Hale, the proprietor of the *Journal of Commerce* who owned the Tabernacle, laid the matter before him, but said I had no money."

That sort of statement ends most discussions with theater owners. "He finally agreed," Colton continued, "that I might have the Tabernacle one evening for $50 and pay him after the exhibition." The Colton charm, which was said to be formidable, must have worked on Mr. Hale. "I then worked the thing up for about three weeks," Colton said. He bought newspaper advertisements that dwarfed those of competing entertainments. The first part of his ad was ballyhoo . . .

> A GRAND EXHIBITION of the effects produced by inhaling Nitrous Oxide, or EXHILARATING OR LAUGHING GAS, will be given in the Broadway Tabernacle, Tuesday Evening March 19 [1844]. SEVENTY-FOUR GALLONS OF GAS will be made so that all in the audience who wish can have an opportunity to take it. THIRTY YOUNG MEN

> have volunteered to take the Gas, so that there may be no delay in case there should be any hesitation among the audience to come forward. TWO GAS BAGS will be used, so that as soon as the effects have ceased upon one person, another bag may be ready for the next. TWELVE STOUT MEN are engaged to stand upon the stage to prevent those who take the Gas[,] while under its influence, from injuring themselves or others.

The second part of the ad was a long quote from *Hooper's Medical Dictionary*, relating Humphry Davy's experience with nitrous oxide. It didn't show the same outright salesmanship as the first paragraph, but Colton considered himself only secondarily an impresario: First he was an educator.

Throughout most of the history of Broadway entertainment, of course, educators have been notably scarce. Not in the early 1840s, however. "Lectures are all the vogue," moaned Philip Hone, a New York socialite, in his diary entry for January 9, 1841, "and the theaters are flat on their backs."

As a blossoming industry, lecturing was part of something still very new at the time: nothing less than the democratization of knowledge. Throughout previous eras, education had been left behind with school: A very few might seek out books and journals, which were quite expensive, but most people contented themselves with the current events found in the newspaper, acquiring little new knowledge on literature or science. Perhaps there didn't seem to be the need, for a person nestled into a proscribed way of life. Perhaps there wasn't the time, anyway, in eras when there was a prodigious amount of daily work for anyone in the middle, lower, and certainly the agriculture classes. However, an urgent interest in self-improvement developed in the 1820s and crested in the 1840s. The first generation born and raised under U.S. democracy took hold of opportunity as a personal reflection of that democracy. What could the great new republics of France and the United States—and the tacitly evolving one in Britain—do for their post-Revolution citizens? Where the ideals had already been won, personal ambition became the new manner of struggle.

By the 1820s people with energy were recognizing with a sudden jolt that energy alone was not going to suffice in their times. They required knowledge—and they knew just that much. Lecturing started in the working class as a thirst for new knowledge. In organizing lecture series, known as "lyceums," by themselves and for themselves, people boldly moved out of the lower or middle classes to which they had been assigned by birth, and placed themselves in a new category, assigned only by aspiration. Nowhere on earth was humanity changing so quickly in the first decades of the nineteenth century as in the United States, and so it is no wonder at all that it was in the United States that the discovery of anesthesia finally occurred.

At the very top, where Gardner Colton proposed to enter the profession, lecturing was a fantastic moneymaking proposition. The circuit made very wealthy celebrities out of favorites such as Horace Mann, Oliver Wendell Holmes, and Henry Ward Beecher, as well as Benjamin Silliman, the chemistry professor from Yale. The reigning king of lecturers, however, was Ralph Waldo Emerson, the former schoolteacher, who not only kindled a reflective turn of mind in his listeners, but many a crush among the schoolgirls who flocked to hear him. Famous personalities represented one type of draw at the top of the lecture circuit. Another was marked by the content, which often succeeded in making education seem as much as possible like entertainment, in the manner of television documentaries today. The vast majority of such lectures covered worthwhile subjects; but it was a balancing act, and sometimes lectures leaned too far toward material of questionable benefit. A lecture offered in 1847 on the "Philosophy and Physiology of the Origin of Life," may have sounded high-minded, but what it looked like was sixteen naked mannequins, representing both genders "at every stage of development." Women and children weren't even allowed in.

Pushing science in another direction, a man named Dr. A. J.

Watts toured the country giving a creditable lecture on electromagnetism. Most of it was creditable, anyway. "Among other wonderful feats, the Dr. will, by particular request, perform the very interesting experiment of causing dead frogs to speak in an unknown and peculiar language": So promised his advertising.

For the Grand Exhibition of Nitrous Oxide on Broadway, Gardner Colton sold tickets at twenty-five cents each, a top price in 1844. Advance sales were tepid, yet on the night of the show there was a crush at the door, as patrons tried to pay their quarter admission directly to the ticket takers. The whole show had to be delayed while Colton sorted the latecomers, their money, and his ticket takers into separate piles.

Finally Colton took the stage and began his lecture. He explained the history of nitrous oxide as a recognized gas, and how it had first been produced about seventy years before by Joseph Priestley. Nitrous oxide is a transparent and colorless gas, and Colton demonstrated how it is made. The common method was to pour a mixture of two parts water and one part nitric acid over copper or iron filings in a closed container. The liquid turns color and bubbles, while nitrous oxide gas is given off and can be collected. Next Colton inhaled some of the gas himself, to show that it was harmless. Finally he offered it to all comers, one at a time.

"Well, the affair came off and my receipts were $535!" Colton exclaimed, looking back on his debut as a lecturer. "I cleared over $400 above all my expenses." Two weeks later he repeated his success. The Second Grand Exhibition closed with an extra experiment, which must have been ever so impressive. Colton described it as: "exploding Oxygen and Hydrogen Gas in a cannon prepared for the purpose—proving that water (which is composed of Oxygen and Hydrogen), although regarded as the opposite of fire, is composed of the most combustible materials."

Before the smell of burning water had even faded from the stage, Colton had enough money with which to continue his medical studies. But that was not to be. "The success," he wrote, "determined me to go on in the business. I gave exhibitions that summer in all the principal towns and cities of New England."

In December, Gardner Colton's New England tour landed in Hartford, Connecticut, where the Grand Exhibition was to be staged at the Union Hall, downtown.

In Hartford, Colton's advertisements attracted the attention of Horace Wells. A dentist by profession, he was a young man of his era, pulled by every breeze within it. Wells took so many things seriously that it is sometimes painful to see the disappointments he bore in return. He was unusual in that respect, Horace Wells, for he would even take laughing gas seriously.

4

THE ENTERTAINMENT IS SCIENTIFIC

I am here as happy as a *clam*," Horace Wells wrote to his sister from Hartford in 1836, using all the current slang, "firing away at teeth, but the greatest wonder is that I have not got on to some other business before this time; or moved to some other place; for I have been here six months—that beats all water."

Raised in comfortable circumstances in Vermont, Wells had moved to Hartford at the age of twenty-one in 1836, not long after taking his training as a dentist in Boston. At the time an apprenticeship of some kind was the only requirement for practice. Wells and the upstanding Connecticut capital city found much to appreciate in each other, and within months he was making enough money in Hartford to worry privately that prosperity would distort his strong religious convictions. Yet Wells didn't foresee continuing in dentistry. He had ideas about seeking his fortune overseas, plans that were discouraged by his mother, who was by then living with her very wealthy second husband in New Hampshire. Instead Horace started a publishing venture in Hartford in the autumn of 1836. He issued a few booklets, and also authored a short book on dental care. In November he wrote to his parents, admitting that he was not quite sure what he wanted to do next. "It is impossible for me to give you any decisive answer now," he wrote about his future

Horace Wells and Elizabeth Wells—miniature oil paintings, circa 1839.
(W. Harry Archer Collection, University of Pittsburgh School of Dentistry)

plans. "It is my sincere desire to do as much good as possible, and I hope and pray that no selfish motive may ever influence me to go contrary to this principle."

That last sentence indicates the endearing, frustrating complexity of Horace Wells, a young man so frenzied with ambition that he could not remain in any one pursuit long enough to profit by it—and yet whose only real motivation was not profit at all, but "to do as much good as possible."

A few years after moving to Hartford, he initiated a formal courtship with Elizabeth Wales, a woman from his church, whom he wed in 1838. The following year they had their only child, a son named Charles. "Our Charley talks and is more interesting than ever," Horace wrote to his mother when the boy was three, "I will give you an example of his intellectual powers—the other day when it was raining he saw a bird on the tree, he says 'Mother poor bird get cold and cough.'"

Rather youthful in his enthusiasms, Horace Wells had a boyish countenance to match. He had a full face, well-scrubbed and rosy, with fine, dark hair that he wore long enough to bloom into tight curls over his ears. From photographs he appears to have been plump as a young man of twenty-four, but he lost weight as he neared thirty, taking on a lean appearance.

"Very appreciative of words of encouragement, he was also very sensitive to criticism," a friend in Hartford wrote of Wells. "Honest himself, he could not think others dishonest; just, he could not brook injustice."

While Horace Wells was building his practice in Hartford, he took in a number of students, several of whom showed genuine promise. C. A. Kingsbury would later help to launch the Philadelphia Dental College, but Wells's star pupil was John M. Riggs, who became Hartford's most prominent dentist throughout the middle of the nineteenth century. He gained international renown after he described the gum disease named for him, Riggs' Disease, or pyorrhea. His work laid the foundations for periodontics, the study of the tissue surrounding the teeth.

While on rounds to outlying communities in early 1842, Horace Wells stopped in Charlton, a town in central Massachusetts. There he crossed paths with William T. G. Morton, already a tired adventurer at twenty-three, recently returned from travels in the West and South. The two became better acquainted, and later that year, Morton moved to Hartford to study dentistry with Dr. Wells, as had Kingsbury and Riggs before him.

Wells taught William Morton the basic practice of dentistry and then helped him establish a practice of his own in Farmington, a village that is today a suburb of Hartford. Ever striving, Wells had several inventions to his credit by that time, including one for a coal screen that received a U.S. patent. Inevitably he originated improvements in dentistry, too, which was one reason that he became so popular with his patients, including among them the governor of Connecticut.

One of Wells's inventions, for a type of gold platework, struck William Morton as especially promising, and in mid–1843 the two dentists formed a business partnership to promote it. The plan was for Wells to help Morton open an office in Boston and then return to Hartford, so that they would have two important cities covered. The partners traveled to Boston together in

October, leasing an office at a prominent address on Tremont Street. Since neither of them had any reputation in Boston, they needed some sort of endorsement. And so they bought one.

And they bought the best: On October 26, the two dentists appeared at the laboratory of Dr. Charles Jackson, located on Somerset Street. Jackson, trained as a physician, was considered the most brilliant chemist and geologist in the region. Morton and Wells agreed to give Jackson twenty dollars for an analysis of the materials used in their plates, and for an opinion of the overall construction. When Dr. Jackson supplied the endorsement, just as hoped, it was a heady moment.

"We have succeeded thus far in our preliminary arrangements beyond our most sanguine expectations," Wells wrote to his wife.

Four weeks later he was trying to extricate himself from the partnership with Morton. Morton had spent the seed money—most of which was Wells's—on himself, without fulfilling his obligation to promote the plates. Wells bore the distress for a year and then, on October 18, 1844, Boston newspapers carried a notice that the partnership of Wells & Morton was formally dissolved. That may have been what the paper said, but Morton still owed Wells books, equipment, plates, and even teeth, in addition to money. Nevertheless Wells continued to regard William Morton as a friend.

Horace Wells was either too good or too callow to harbor the proper bitterness toward a scoundrel. Or too full of brotherly love and optimism: Whatever his very unique constitution, it was not to be regretted. It was the first stroke in an unusual convergence of men—a convergence that was to beat sheer genius to the discovery of surgical anesthesia.

While continuing in dentistry, Wells was still distracted by his nagging sense of ambition, and he spent much of his free time tinkering with inventions and ideas that excited his imagination and trying to gather knowledge around him, in the fashion of his day. In July 1844, he and Elizabeth attended a demonstration of animal magnetism, another name for hypnotism. In the exhibition a woman named Mrs. Powers went into a trance and then told each of the Wellses what parts of their bodies were ailing. Horace proclaimed her to be fantastic and her conclusions to be amazingly accurate. He even listened to her lecture on the subject of light, spiritual and natural, which was delivered while she was in a trance. He thought that was terrific, too.

In the second week of December 1844, a little more than a month after Horace Wells formally dissolved his partnership with William Morton, Gardner Q. Colton came to Hartford. He busied himself with arrangements for a "Grand Exhibition of the Effects Produced by Inhaling Nitrous Oxide." The advertisement Colton placed in the *Hartford Courant* was his usual boldface ballyhoo, in gigantic print, promising FORTY GALLONS OF GAS, referring to those STOUT MEN who would be on hand to render control, and providing that attractive double enticement of silliness and self-improvement. "The entertainment is *scientific* to those who *make* it scientific," he declared, immediately after predicting that he would make those in the audience laugh more than they had for the previous six months. "Entertainment to commence at 7 o'clock, at Union Hall. Tickets 25 cents," the ad concluded.

Horace and Elizabeth Wells were among the many who held tickets. After dinner on Tuesday, December 10, they journeyed downtown. The evening should have been nothing more than a fleeting diversion. Instead Mrs. Wells would later look back on it as the beginning of "an unspeakable evil."

Henry Wood Erving, a longtime resident of Hartford, liked to attend Grand Exhibitions whenever Mr. Colton was in town. "The gas used in these lectures by Dr. Colton was contained in a rubber bag," he recalled,

> and was administered through a horrible wooden faucet, similar to the contraptions used in country cider barrels. It was given in quantities only sufficient to exhilarate or stimulate the subjects, and reacted upon them in divers and sundry ways. . . . Dr. Colton, after a short lecture regarding

Cartoon drawn in 1808, spoofing the nitrous oxide fad.
(Edgar Fahs Smith Collection, University of Pennsylvania)

> the nature and properties of the gas, nearly always took the first dose himself—self-administered—declaiming quite wonderfully afterwards, and invariably winding up with his hand to his head and the remark, "The effect now is nearly gone."

It certainly didn't require an inhalation of nitrous oxide to induce Gardner Colton to declaim: a whiff of common air and the drop of a hat were all the encouragement he needed to quote a whole act or two of Shakespeare, but his evident survival impressed the audience that the gas was harmless. A small stampede took to the stage when he asked for volunteers in Hartford.

Dr. Horace Wells was in the throng. Among the others in the first group of ten chosen to try the gas were two people he knew quite well, David Clarke, a friend, and Sam Cooley, an acquaintance who worked as a clerk at a drugstore owned by Abiel A. Cooley, a relative. Wells took his turn at the bag and duly made a fool of himself, according to his wife. She never quite specified how, and neither did he. However, he did relate in great detail what happened immediately after his gambol with laughing gas, when he was coming to his senses again, in the row of chairs at the back of the stage. Wells looked around and noticed that Sam Cooley, sitting right next to him, had blood on the knees of his pants. Wells probably wasn't aware that a few minutes before, Cooley had been dancing and hopping all over the stage, and that, to the delight of the audience, he had plowed straight into a wooden settee on the stage.

"You must have hurt yourself," Wells said to him, as they each came to. Cooley had only just begun to feel his knees smarting. He looked down at his legs and was surprised to see just how badly he had been hurt. He told Wells that he hadn't felt anything while he was affected by the nitrous oxide.

Wells thought about that for a while.

Then he told his friend David Clarke that a person could probably "have a tooth extracted or a limb amputated and not feel any pain." Even after he had returned to his seat in the audience, Wells was still thinking it through. When Colton's show was over, Dr. Wells returned to the stage and approached Gardner Colton for a second time.

"Why cannot a man have a tooth pulled while under the gas and not feel it?" Wells asked, posing the most important question of his whole career.

"I don't know," Colton replied. No one really knew the answer to that question. Colton did Wells a service by refraining from a repetition of the old line, the one repeated in books, that nitrous oxide had no serious uses and that it was a fatal poison when taken in large doses. With Colton's honest ignorance as a kind of encouragement, Horace Wells left the theater and went directly to see his friend and former student John Riggs. As the hour grew late, they discussed the idea of using nitrous oxide gas to produce senselessness and made the decision to try an experiment the very next morning, while Gardner Colton was still in town.

Because it was Horace Wells's idea he had to go first, according to the code of valor accepted by scientists at the time. He was to inhale the gas in a greater dosage than he had the previous night, and in a far more serious frame of mind—that being a guiding factor in the effect of the gas, whether toward sedation or exhilaration. Once the gas took its effect, Wells would have a tooth extracted by Dr. Riggs.

"We knew not whether death or success confronted us," Riggs said, "It was *terra incognita** we were bound to explore."

The next morning Wells went back to Union Hall, found Colton, and asked him for a bag of nitrous oxide. Colton suggested that he himself should be on hand, and so the two of them left the theater for 180½ Main Street, where both Riggs and Wells had their offices. Somehow, along the way, Sam Cooley and two other men joined the group. In Wells's office the three spectators positioned themselves by the door, ready to run when the patient became violent, which they were sure he would do. Gardner Colton set up his equipment—the bag of nitrous oxide and the "horrible wooden faucet"—next to the dental chair, and then he, too, stood back near the door. Only Dr. Riggs remained nearby, ready to make the extraction the very instant the gas took effect; the tooth selected was one that had bothered Wells for some time, a stubborn wisdom tooth that would certainly test any painkiller. When each of the other men was ready, Horace Wells sat himself down in the chair and without any further fuss, anesthetized himself, breathing calmly and deeply of nitrous oxide.

"Wells took the bag in his lap," Riggs wrote, "held the tube to his mouth & inhaled till insensibility relaxed the muscles of his arms—his hands fell on his breast—his head dropped on the

---

*Unknown territory.

head-rest & I instantly passed the forceps into the mouth, onto the tooth and extracted it."

Within minutes Wells awoke and felt the space in his mouth, where the tooth had been. "It is the greatest discovery ever made," he cried. "I didn't feel it so much as the prick of a pin!"

Wells and Riggs repeated the experiment on others, and word spread throughout Hartford that there was something wonderful going on at the dental offices on Main Street. One patient told another, and before the month of December was over, more than a dozen people had experienced Wells's new alternative to pain. His office had never been busier. Late one afternoon he even told those still waiting that he was "tired and lame" from standing in his office all day long, administering gas and pulling teeth.

The new discovery was anything but a secret. Dr. Wells was far too enthusiastic a man to keep one; he lost no time in calling on each of the other dentists in Hartford to explain the use of nitrous oxide. Eventually several learned how to administer it to their own patients.

To their lasting honor, neither Gardner Colton nor Dr. John Riggs ever tried to claim even partial credit for Horace Wells's discovery. Wells himself seems to have been a bit ashamed of the circumstances that triggered his idea—the entertainment lecture and his appearance onstage—and he later claimed to have deduced the discovery while "reasoning from analogy." He wasn't reasoning from analogy, though. He was staring right at it, in the form of Sam Cooley's bloodied knees.

Sam Cooley's legs may have been asleep, but his ears were wide awake by the time Wells began his first experiment on December 11. That morning, while Horace Wells was gathering Gardner Colton and John Riggs in his office, Cooley—the druggist's assistant—was at another dental office, holding a bag of his own nitrous oxide and requesting cooperation in an experiment with it. The dentist refused, having read one of those many books that warned of the dangers of nitrous oxide. Only then did Cooley join the group in Riggs's office.

Horace Wells may have been too ebullient to keep his discovery a secret, but he was far too ambitious to ignore all its possibilities. "Wells and myself entered into a verbal agreement," Sam Cooley later claimed, ". . . with the intention to establish an office in New York and Boston, one of us to reside in each of those places."

"Our plan," Cooley continued, "was to manufacture the gas ourselves which was to be kept in a gasometer in an adjoining room with a pipe leading from it to the operating chair. By this means we expected to keep the whole matter a secret and under our sole control, which would insure us a large and lucrative business." That was Cooley's story later, though it is hard to think how Wells could be making plots about secrecy and pipes and very large fortunes with Sam Cooley, when he was at the same time blabbing all over town about painless dentistry and nitrous oxide.

As soon as all the dentists in Hartford knew about the discovery, Wells made plans to go to Boston with it. He was ready, whether the discovery was or not.

In late January, Wells contacted his erstwhile partner William Morton, who in turn contacted Dr. John C. Warren, head of surgery at Massachusetts General Hospital. Dr. Warren was willing to meet Wells and so give an audience to the new discovery.

At the first meeting later in January 1845, Dr. Warren and the house surgeon, George Hayward, treated the Hartford dentist with "much kindness and attention," in Wells's own recollection, and Dr. Warren was so satisfied with the potential of nitrous oxide that he invited Wells to lecture the medical students about it. In the meantime Wells paid a social call: "While in Boston," he wrote, "I conversed with Drs. Charles T. Jackson and W.T.G. Morton upon the subject, both of whom admitted it to be entirely new to them." Wells also used his free time to supply nitrous oxide for the amusement of the students. He may have thought that he was raising their understanding of its properties, but they were not used to regarding it as a serious scientific substance. Wells's glad efforts in dispensing laughing gas probably only lowered their opinion of the man making claims for it.

"Dr. Wells was introduced to our class by Dr. John C. Warren, then Professor of Anatomy at the University," recalled a former medical student at Harvard, C. A. Taft, "Dr. Wells then made a statement of his discovery, spoke of its importance, and his hopes of introducing it—the anesthetic agent—into general use in surgical operations."

Dr. Warren authorized Wells to administer the gas to a patient scheduled for an amputation. After delaying for a few days, however, the patient retreated home, too frightened to undergo an operation. No other operations were scheduled for the week, and so a case of elective surgery had to be solicited. "At this time," Taft explained, "Dr. Wells extracted a tooth for some one under the influence of the gas." The patient was a student who volunteered for the experiment as a last-minute substitute for the scheduled operation. A number of students and faculty members came to observe, as did William Morton, who brought dental tools for Wells's use.

Standing at the patient's side, Dr. Wells administered enough gas to produce a likely stupor. In Hartford, Wells had depended on Sam Cooley for his supply of nitrous oxide, a gas that can be manufactured by any one of several different processes. No matter which is used, the result should be pure nitrous oxide. However, variances in temperatures or in timing can produce a slightly adulterated version of the gas.

Wells—the master with a month and a half of experience under his belt—was at a disadvantage in procuring nitrous oxide in Boston, where the quality may well have been different from that manufactured by Gardner Colton or Sam Cooley. It may even have been better, purer. But in any case Horace Wells would not have known. He only barely knew what he *knew*; he was oblivious to all that he didn't yet know. Nonetheless Dr. Wells was very sure of himself as he set about pulling the tooth of his logy patient. At first all went well, but then there was a noise. Some later called it a groan. Others called it a weak sort of a bleat. Dr. Taft reported: "The patient halloed* somewhat during the operation."

The students in the audience then "halloed" a lot, and they jeered at their laughing-gas man. He heard the word "humbug," called out over and over again. No one took much notice when the patient awoke fully and protested that he had experienced almost no pain.

All the pain in the room was in Horace Wells's body. "As several expressed their opinion that it was a humbug affair (which in fact was all the thanks I got for this gratuitous service)," Wells wrote three years later, "I accordingly left the next morning for home." He didn't show his face but dropped Morton's dental instruments off on the doorstep at daybreak and slunk away. He would later blame the debacle on the fact that he had taken the bag away too soon.

Horace Wells did not earn his great discovery in the obvious way, by studying medicine in order to understand pain or the sciences to know chemistry. Wells's painstaking effort was, if anything, one of unblinking desire: the way he lived his life and dreamed his hours away, trying to imagine a great improvement to which he could sign his name. Having found it, though, he also found out just how powerless he was.

"The excitement of this adventure," Wells said later, "brought on an illness from which I did not recover for many months."

---

*Cried out.

5

UNWILLING COLLABORATION

Horace Wells often looked back on his meeting with Charles Jackson and William Morton in January 1845, just before the failed demonstration at Mass. General, and on that moment when he described for them his use of nitrous oxide in dental surgery. "Dr. Jackson particularly seemed inclined to ridicule the whole thing," Wells remembered.

A Hartford physician named Dr. E. E. Marcy, who visited with Horace Wells right after his return from Boston, acknowledged that that had also been the impression Wells made at the time. "He informed me," said Dr. Marcy, "that his discovery and his whole idea respecting anaesthetic agents was ridiculed by Dr. Jackson and other medical men of Boston."

Then Marcy added another fact concerning Wells's impression of his meeting with Jackson and Morton—Jackson and the other medical men may have ridiculed the discovery, but "his former pupil, Morton, swallowed this ridiculous idea greedily."

William Morton knew, as did no one else in Boston, that Horace Wells was anything but a "humbug": that he was an original thinker and a highly competent man.

A half year later, in July 1845, Morton paid a visit to Horace Wells in Hartford. Officially he was there in answer to Wells's plaintive requests for some of the money still owing from the days of their partnership. Wells was thrilled and relieved to see his prosperous former partner finally willing to settle old accounts. Morton made at least a start in the reckoning, but that was not the only reason he was at Wells's home.

If Horace Wells had been slightly more suspicious—if he had even one suspicious cell in his whole body—he might have wondered why William Morton was so flush with money and business, when their partnership, only recently ended, had never yielded anything but debt. Wells—who had been rich when he met William Morton—was destitute in the summer of 1845. Meanwhile, that same summer, William Morton—who had been derelict when he met Horace Wells—was in the process of buying a grand estate.

With an upper-middle-class income of between ten and twenty thousand dollars per year, Morton and his wife felt the need of a country property and bought one about eleven miles from Boston in West Needham (now known as Wellesley), then just becoming fashionable as a retreat. By 1845 the couple had been joined by the first of their five children, and they lived in lordly style, with extensive gardens and an enviable social life. Daniel Webster, the senator, had a farm of his own nearby and brought over a pair of Chinese geese as a housewarming gift for Morton. The farm soon gave William T. G. Morton a name as one of the state's most progressive farmers—yet the deed to the farm was in his father's name. On paper W. T. G. Morton had no assets—at least he made sure that no one really knew what he had.

None of the account books survive from the Wells-Morton partnership, but some time later Charles Jackson noticed something about Morton's dental practice in Boston that probably applied to the defunct partnership of Wells & Morton. "I discovered," he said, "that Morton did not pay his debts, but kept steadily investing his money *under cover* of the name of a Mr. Whitman* of Connecticut [emphasis in original]."

---

*William Morton's in-laws were named Whitman and were from Connecticut.

"I did not then know of his Western career as a swindler," Jackson added. Neither did anyone else, as Morton embarked on his more furtive Eastern career as a swindler.

While William Morton was at Wells's home in Hartford, going over old bookkeeping and making small talk, he remembered to ask Wells for more information about his use of nitrous oxide gas to produce insensibility. At the time Wells was not practicing dentistry himself, but he regularly administered the gas for Dr. John Riggs, who was using it about once a week on his patients. Wells was happy to bring his old friend in to observe a case of painless dentistry at Riggs's office. For his part Morton seemed to be fascinated, although he did nothing about the discovery for about a year.

Morton later claimed that he had long been on the lookout for a painkiller of some sort, simply to expedite his growing dental business. He told stories, no doubt true, of patients who needed to have a tooth pulled, coming into the office and simply sitting in the chair, while they tried to summon the courage to let Morton go ahead. Patients sat in Morton's chair for hours on end. It was hard to make money when patients came into the office only to sit in the chair and then go home.

In Hartford they wouldn't have had cause to fear. A number of patients there were accustomed to painless dental surgery in 1845–46. Elizabeth Williams, a Hartford resident, took advantage of the new technique and inhaled Dr. Wells's nitrous oxide before having a tooth pulled in John Riggs's office on March 6, 1845. The following summer, in 1846, she happened to be vacationing at a town called Stafford Springs in Connecticut, when she met the personable William Morton, also on holiday there. "Learning that he was a dentist, I spoke of my tooth, and mentioned the fact that Dr. Wells had administered gas to me," she said. "I remarked to him that I was among the first that took the gas. He asked about the effect and operation of the gas, and made no intimation of any acquaintance with or knowledge of the gas."

Morton's ignorance must have been feigned, for conversational if not more darkly motivated reasons. In any case it is hard to believe that he could have forgotten all about the experiments with nitrous oxide that he'd witnessed in Boston and Hartford. Undoubtedly, though, Elizabeth Williams's story reignited William Morton's interest in Wells's discoveries regarding nitrous oxide. It did not concurrently reignite any inclination to seek out Wells again. By the summer of 1846, Morton was avoiding Wells entirely and ignoring his former partner's requests for some settlement of their finances. Yet it was sometime that summer that he began acting upon his interest in a medical painkiller.

For his part, Morton insisted that the idea for a surgical painkiller had not risen at all out of his association with Wells. According to his version of the discovery of anesthetics, he was at his office one day when it occurred. He was, so he said, listening to a new assistant describe the ether frolics staged at his old school, when he realized that ether could serve as the painkiller he had so long required in his dental business.

The next time Morton was at his farm in Wellesley, he said he experimented with ether, using for a subject his dog—his father's spaniel, to be accurate. The dog was named Nig. Morton pushed Nig's head into a jar partly filled with sulfuric ether, and noted how the dog then drooped in his arms for about three minutes. In early August, Morton decided to try the experiment again, but Nig took one look at the jar and declined to participate. In fact he broke the jar on his way out of the room. Morton sopped up the ether that had spilled on the floor and inhaled it from his handkerchief. In the blear that followed, he concluded that he would not have felt any pain had a tooth been extracted at the time.

Looking around the farm, Morton saw a whole laboratory of animals on which to experiment—he thought the same thing in looking around his office in Boston and seeing his assistants. However, Morton had a problem that only a thinking opportunist would have: If he ordered ether from the same druggist for a second or even third time, someone there might wonder what was going on and start thinking about sulfuric ether and its possible uses in dentistry. Sneak that he was when he needed to be, he sent a different assistant to a different drugstore everytime he wanted to purchase more ether. With each new supply he etherized more animals on the farm, including bugs by the jarful and even big worms, the green ones that crawled around on his grapevines. First, of course, he had to determine the difference between a wide-awake big green worm and one that was unconscious.

"Do you put the fish asleep, too?" Morton's wife once asked.

"I try to," he replied in a serious tone, "but I have not succeeded yet."

Some people, as a matter of fact, doubted that Morton had actually undertaken any animal testing on the farm. Charles Jackson found something to snicker about in one of Morton's early descriptions of Nig's reaction to ether—jumping ten feet in the air and then running away, only to fall into a pond. A dog, Jackson noted dourly, could not even stand up under the influence of ether, let alone jump ten feet. "I would not give the snap of my finger for Dr. Morton's alleged previous experiments," Dr. Augustus Gould agreed, "except so far as they go to show his having been previously considering the subject." Precisely what subject Morton was indeed mulling over is not clear, but it is a matter of record that late in the summer of 1846 he went to Joseph Wightman's workshop to hunt around for some means of administering a gas. Wightman suggested that Morton ask his own mentor, Charles Jackson, for a suitable gasbag.

"I approved of the suggestion, but feared that Dr. Jackson might guess what I was experimenting upon," Morton wrote, allowing that "by this admission, I may show myself not to have been possessed by the most disinterested spirit of philosophic enthusiasm, clear of all regard for personal rights or benefits."

"But it is enough for me," Morton continued, "to say that I felt I had made sacrifices and run risks for this object, that I believed myself to be close upon it, yet where another, with better opportunities for experimenting, availing himself of my hints and labors, might take the prize from my grasp."

It is hard to think what sacrifices William Morton had made during three or four weeks of etherizing his pets and the local insects. On the basis of risks Nig had more claim to glory than did William Morton. Time, sacrifices, and risks had nothing to do with it. With the answer in plain sight, the first one to solve the problem of surgical pain was going to be a man on a mad dash. Morton may have been close to the breakthrough, but having no knowledge of chemistry, he needed instant expertise and knew where to get it.

William Morton later savored the recollection of fooling Dr. Jackson into a short lecture on sulfuric ether on the morning of September 30. "I asked Dr. Jackson for his gas bag," Morton wrote.

> He told me it was in his house. I went for it, and returned through the laboratory. He said, in a laughing manner, "Well, Doctor, you seem to be all equipped, minus the gas." I replied, in the same manner, that perhaps there would be no need of having any gas, if the person who took it could only be made to believe there was gas in it, and alluded to the story of the man who died from being made to believe that he was bleeding to death, there being in fact nothing but water trickled upon his leg; but I had no intention whatever of trying such a trick. He smiled and said that was a good story, but added, in a graver manner, that I had better not attempt such an experiment, lest I should be set down as a greater humbug than Wells was with his nitrous oxide gas.

"Seeing that here was an opportunity to open the subject," Morton recalled, "I said, in as careless a manner as I could assume, why cannot I give the Ether gas?" (Morton's terminology is inaccurate: At room temperature ether is a liquid, and it is the fumes that are inhaled.)

"He said I could do so," Morton said, describing Jackson's response.

> He said the patient would be dull and stupefied, that I could do what I pleased with him, that he would not be able to help himself. Finding the subject open, I made the inquiries I wished as to the different kinds and preparations of Ether. He told me something about the preparations, and thinking that if he had any it would be of the purest kind, I asked him to let me see his. He did so . . . Dr. Jackson followed me to the door, and told me that he could recommend something better than the gas-bag to administer the Ether with, and gave me a flask with a glass tube inserted in it.

Charles Jackson's lawyers entirely refuted the Morton story about what happened on the morning of September 30. According to them:

> Mr. W. T. G. Morton called at the office of Dr. Jackson and requested the loan of an India rubber bag, for the purpose of administering atmospheric air to a patient, in order to act upon her imagination, and to induce her to permit him to extract a tooth. He was dissuaded from the attempt by Dr. Jackson. There was also some conversation concerning nitrous oxide.

"It was finished," the lawyers' report continued,

> and no request had been made by Mr. Morton to Dr. Jackson to suggest to him any process by which teeth might be extracted without pain. Mr. Morton had left the apparatus room, in which most of this conversation had occurred; he went into the office on his way to the street, when he was followed by Dr. Jackson, and stopped by him. Dr. Jackson then informed Mr. Morton that he could impart to him a means of producing a general insensibility, during which, he was confident, surgical operations might be performed without pain. He communicated to

> Mr. Morton all that it was necessary he should know, for the performance of this experiment. He informed him that the substance to be used was the vapor of sulphuric Ether . . . . He gave instructions in the most minute detail; so that nothing whatever was left for Mr. Morton to devise in any part of the process.

According to the report made on behalf of Jackson, Morton's uninformed response was, "Is it a gas?" Dr. Jackson's contention was that William Morton had never heard of sulfuric ether before their meeting on September 30.

Bolstered by Jackson's briefing, however it may have been elicited, Morton raced back to the office and looked for a dental patient willing to participate in a trial of ether fumes. None came forward, despite the five-dollar incentive Morton's assistants offered to passersby on Tremont Row. Mrs. Morton recalled that her husband finally tested it on himself on that day, September 30. Morton left a vivid description of his sensations under the influence of sulfuric ether, which he inhaled from a handkerchief.

"I looked at my watch and soon lost consciousness," Morton wrote.

> As I recovered, I felt a numbness in my limbs, with a sensation like a nightmare, and would have given the world for some one to come and arouse me. I thought for a moment I should die in that state, and that the world would only pity or ridicule my folly. At length I felt a slight tingling of the blood in the end of my third finger and made an effort to touch it with my thumb, but there seemed to be no sensation. I gradually raised my arm and pinched my thigh, but I could see that sensation was imperfect. I attempted to rise from my chair, but fell back. Gradually, I regained power over my limbs and full consciousness. I immediately looked at my watch and found that I had been insensible between seven and eight minutes.

William Morton considered the experiment a complete success and he was anxious to test ether truly—in an actual dental procedure. Mrs. Morton reported that her husband came home at dinnertime, protesting that "late as it was, he must still find a patient." Back at the office all evening, though, he couldn't find anyone desperate enough to let him try the new discovery.

"He was on the point of Etherizing himself once more," Mrs. Morton continued, repeating the family lore, "and having one of his assistants extract a tooth from his own head, when there came a faint ringing at the bell." At the door was a man with a smarting tooth; he asked Morton to extract it, but begged to be mesmerized first.

"The doctor could almost have shouted with delight," Mrs. Morton reported, meaning by "doctor" her husband. Instead of mesmerism or anything like it, Morton soberly talked the man into inhaling sulfuric ether four times from a handkerchief. Morton's assistant was trembling with fear, so Mrs. Morton said, but more important, the patient was rendered immobile. A few moments later Dr. Morton sprinkled water on the man's face, and he came around.

"Are you ready now to have the tooth out?" Morton asked him.

"I am ready," the patient said.

"Well, it is out now!" Dr. Morton had the great satisfaction of declaring, as he pointed to the bicuspid on the floor.

Parts of Morton's version of his discovery were true, and parts were not. For example, his rather gothic description of the tense wait for a patient, any patient, on the night of September 30 was not the whole truth. He didn't mention the fact that the man who arrived for the operation was his good friend, Eben Frost, a wood sawyer by profession. He also refrained from mentioning he had arranged for a newspaperman named Albert Tenney to be present at the emergency extraction. Morton had enlisted Tenney to publicize the wonder of the inhalation without disclosing what it was.

One man who was not present at the Frost etherization was

Dr. Charles Jackson. He was at his home on Sumner Street, only a short way from Morton's office on Tremont Row. And yet, according to the version that Jackson would later relate, the experiment was entirely *his* triumph. He was the puppeteer. and William Morton was merely a device that he brought to life with his genius.

The events of September 30, 1846, were critical to the subsequent lives of Morton and Jackson.

Somehow or other William Morton had emerged from the laboratory of Dr. Charles Jackson on September 30, prepared to try sulfuric ether in an actual dental procedure, which he did, that very evening. Newspapers ran the story of Morton's new painless dental surgery on the day following Eben Frost's etherization, proving that Albert Tenney was on the job. "A New and Valuable Discovery," the *Boston Transcript* called it one day after that, on October 2. The story ended by stating quite correctly, "The discovery is destined to make a great revolution in the arts of surgery and surgical dentistry."

Dr. Henry J. Bigelow, a junior member of the staff of Mass. General, read the article and latched onto the import of its last sentence, as no one else did. Seeking out William Morton, he observed thirty-seven trials of ether at the dentist's office during the early part of October. By October 4 Morton was advertising "painless dentistry" in bold type in the Boston newspapers, attracting a clientele that kept him and his assistants busier than they had ever been before. It was then that William Morton decided he'd better find out even more about ether. He did so by sending his wife's brother, Francis Whitman, to ask Charles Jackson for some chemistry books.

"I went to Dr. Jackson's and he spoke to me of some notices in the papers," recalled Whitman, making reference to Morton's advertising campaign, "but immediately after, said he did not 'care how much Dr. M. advertised, if his own name was not drawn in with it.'"

While the haughty Charles Jackson steered clear of William Morton and his mercantile dentist's office, twenty-eight-year-old Henry Bigelow could hardly stay away.

• • •

There is a certain similarity in the experiences of Horace Wells and William Morton in discerning the efficacy of inhalation anesthesia (nitrous oxide in Wells's case and ether in Morton's) and then rushing with it to Mass General, but a fundamental difference between the two lay in the form of Henry J. Bigelow. The young surgeon paved a path for Morton and, what is more, made sure that Morton stayed on it. One by one Bigelow brought his colleagues from Mass General's surgical staff to witness the etherizations in Morton's office. Bigelow's growing role as an intermediary between Morton and the hospital gave him a stake in the success of the discovery. Horace Wells, two years before, had had no one but himself.

Morton also had something else that no serious medical researcher had ever had before, not Wells, certainly, but not Jenner, Davy, or the great Larrey either. Morton had a patent attorney. On October 1, the day after his first successful etherization, he was already conferring with R. H. Eddy, Esq., regarding the possibility of a patent on ether. No one had ever patented a truly important medical discovery before. Quack doctors patented elixirs and gadgets, of course, taking up pages of newsprint in every city in the country to advertise them.

A self-respecting doctor would *never* patent *any* breakthrough that might reduce human suffering. Most doctors regarded as redundent the bylaws in their state medical associations that expressly prohibited members from hoarding "secret remedies." No good doctor would. Perhaps they were not all saints, the doctors of the 1840s, but they definitely weren't businessmen.

At the meeting on October 1, Eddy said that he would look into the patent possibilities for Morton. And the possibilities that floated through Morton's imagination must have been staggering. Control over the use of anesthesia would give him the right to exact a fee for every surgical operation performed anywhere. But that was peanuts: It would also give him a slice of every dental practice in the whole country. The numbers only multiplied when he considered overseas countries. Morton and Eddy later calculated that the very smallest amount that could be realized over the fourteen-year life of a U.S. patent was $365,000 (equivalent to about twenty times that today, or at least $7 million). Foreign licenses and other secondary sources (such as equipment sales) would add to the initial figure.

While Eddy was looking into the law, Morton was experimenting further with the discovery. First, he learned to add oil of orange to the ether, in order to disguise the signature scent of the chemical; the addition also gave him the right to assert that the secret liquid was not merely one substance, but a compound. Second, he looked for ways to augment the fortune to be made on ether, accelerating his need for a special apparatus with which to administer it.

On October 15, the day that Morton received permission to demonstrate "the preparation" at Mass General, he went into a panic as he tried to fashion in a matter of hours a new apparatus to use in the public demonstration. If the safety of the patient were his main concern at that time, he would have used a cloth to administer the vapors, as he had many times before. His concern, however, was the viability of the discovery as a commercial product, and so it was that he rushed his friend Dr. Gould and two different instrument makers through a relay race of ingenuity and construction.

Morton had to turn a discovery into an *invention.* He had less than twenty-four hours in which to do it.

6

SILENCE IN THE DOME

William Morton strode into the operating theater at Mass General at twenty-five minutes after ten on October 16, 1846, carrying his apparatus with him. John C. Warren stepped back from the patient and received Morton into the small circle surrounding the operating chair. "Well, sir," Dr. Warren said to him, "your patient is ready." His voice carried a sarcastic edge that drew a snicker from the audience.

To Morton the next few minutes were probably a blur; his only recollection of them was that "[t]here was a full attendance, the interest excited was intense." But that didn't mean he'd lost his self-assurance. As he set up his apparatus—the glass retort with two openings, a tube extending from one of them—he explained to Dr. Warren why he'd been delayed, overseeing last-minute adjustments to his equipment. He unscrewed a brass cap from the opening on one side of the bulbous glass retort and pushed his finger through the cork flap just beyond, in order to prop it open. He then poured in enough of his liquid compound to cover the bottom of the glass globe. The cork operated as a valve that was supposed to open in concert with the inhalation of the patient. It was attached to a spring, like a trap door. When Morton removed his finger, the cork flap snapped shut from inside the retort. Morton was familiar with that part of the apparatus, and his hands, angular and well-defined, were probably impressive in going about the filling process. However, he had never operated the valve system just fashioned by Chamberlain. According to one version of the story, Morton had been waiting outside the hospital door since before ten o'clock; Chamberlain's messenger had delivered the improved retort to him there. He had only as much experience with the valves as he could absorb as he hurried up the stairs to the operating theater.

The valve system, contributed by Dr. A. A. Gould at the eleventh hour—quite literally the evening before—was designed to keep exhaled air from reentering the glass globe. In addition to the side valve through which Morton had poured the liquid, the tube that led to the patient's mouth from the other side of the apparatus featured two more valves. They were also covered by cork flaps, operated on springs—two more little trap doors. One of them, the connection valve, was between the tube and the glass globe. The other was the tube valve, and it was, naturally, on the side of the tube. As the patient inhaled through the mouthpiece at the free end of the tube, the two flaps on either side of the glass globe would swing open with the sucking air. Common air would come in through the side valve, mix with ether fumes, go through the connection valve and then on through the tube to the patient's mouth. With each inhalation, meanwhile, the tube valve would fix itself shut. (It wouldn't do any good to have the ether fumes go out the side of the tube without reaching the patient.) When the patient exhaled, that same tube valve would swing open and let the old air out. The connection valve would then be pushed shut to protect the atmosphere inside. However, it is doubtful that the valves were supple enough to be manipulated by the breathing of the average person, and Morton almost certainly had to use his fingers to regulate them, particularly the first one, the one on the side of the globe. It was responsible for the balance between air and ether fumes. Poor combinations of the two promised momentary or prolonged suffocation of the patient, or etherization insufficient for even a flea.

Turning two valves on a retort, in concert with the breathing of another person, is not necessarily as difficult as playing a full concert on the piano, but William Morton arrived center stage at the operating theater in Mass. General without ever before having touched his particular piano—the improved retort. And so, when he placed the mouthpiece in the patient's lips and told him to breathe deeply and slowly, he didn't really know what on earth he was doing. But he did it anyway. That is the glory of the man. He could have been reviled for the exact same thing, and that is the glory of his age, the impatient 1840s when William T. G. Morton was center stage.

Elizabeth Morton recalled that she stayed home all day on October 16, sitting by the window, "expecting every minute some messenger to tell me that the patient had died under the ether and that the doctor would be held responsible!"

In the operating theater Morton was administering ether in public for the first time. Everyone in the room was watching the unusual process intently. After about three or four minutes, the patient, Gilbert Abbott, became insensible and fell into a deep sleep. Dr. Morton removed the mouthpiece. He stood back, bowing to Dr. Warren.

"*Your* patient is ready, sir," he said, returning the chief surgeon's joust. Dr. Warren delayed no longer but picked up an instrument and proceeded with the operation to remove the tumor from Abbott's neck.

"I immediately made an incision about three inches long through the skin of the neck and began a dissection among the important nerves and blood vessels of the neck," reported Dr. Warren, "without any expression of pain on the part of the patient."

Ether Day had finally arrived.

When the operation was over, Dr. Warren turned to those watching rapt from the steeply banked seating of the operating theater. As if to say aloud what they were all thinking, he announced in a sober tone, "Gentlemen—*this* is no humbug." After a pause his words were greeted by cheering.

Asked repeatedly, Abbott confirmed—repeatedly—that he had felt no pain during the operation, although by the end of it, he said, he became aware of a sensation he compared to the scraping of a blunt tool, such as a hoe, against his skin. Dr. Warren accepted the diminution of the vapor's success on the basis that the procedure had taken longer than the four or five minutes of painlessness Morton had promised in the first place.

"Two o'clock came," Mrs. Morton said of her vigil at home. "[T]hree o'clock and it was not until nearly four that Dr. Morton walked in, with his usually genial face so sad that I felt failure must have come. He took me in his arms, almost fainting as I was, and said tenderly: 'Well, dear, I succeeded.'"

Mrs. Morton described her husband at his moment of triumph as "sick at heart" and "crushed down, one would have said, by a load of discouragement." She was never able to explain why he should have been affected so morosely by the excitement of the day.

Ether Day was Morton's triumph. It was superb. It was immense. It was much too magnificent for one such as he.

7

THE CONFIDENCE MAN

W.T.G. Morton, when he first came to Boston and set up as a Dentist," Dr. Jackson recalled, with the brunt of his later antipathy, "could not write grammatically his own quack advertisements and was in the habit of bringing them to me to write out for him in readable English, before taking them to the newspapers." Augustus Gould and Horace Wells also made note of Morton's inability to write legibly or grammatically.

"On one occasion," Jackson continued, "I urged him to abandon the system of puffs and quack advertisements and told him what course a respectable dentist should take in making himself known to the public. Soon after," Jackson wrote,

> he called upon me and professed to be desirous of becoming a *respectable dentist* and solicited "*the honor*" of entering his name with me as my pupil, preparatory to attending the medical lectures, which I had advised him to follow. He seemed so plausible in his appearance and was apparently so sincere that I offered all proper encouragement, but declined to take him as a medical pupil, for want of time to attend him. He urged me so strongly, declaring that he would not ask any time when I was busy, etc. that I at last consented that he should become my *nominal* pupil.

"He was about to marry a girl in Connecticut . . . ," Dr. Jackson added in passing.

And Morton did marry that girl, Elizabeth Whitman. In the mid-1840s he had a burgeoning dental practice in Boston, an association with a highly esteemed physician, and a sweet wife who was the pride and joy of an old Connecticut family.

To put that another way: William Morton was entered as a student of the eminent scientist, Charles Jackson, despite the fact that he could not for all intents and purposes write, and he became a member of the well-respected Whitman family of Farmington, Connecticut, despite the fact that he was a wanted man in locales from Rochester to New Orleans.

"My mother was a member of the Baptist Church ever since I can remember," William Morton recalled in a letter to a friend. "It was her influence that educated me morally," he added, granting her a rather dubious distinction. William Thomas Green Morton was born August 9, 1819 on a farm in Charlton, Massachusetts. By the time he was a teenager his parents had sold the farm in order to start a store. They soon ruined themselves, though, and William was forced to quit school for a job in a tavern in Worcester, twelve miles from Charlton. After he was caught embezzling money, he left Worcester, in turn, under sheriff's orders never to come back.

William Morton started over many times more than the average person, but then his mistakes disappeared from his mind, long before they had a chance to turn into remorse. So it was that the handsome face with which he greeted each new round of believers was as fresh and earnest as ever it had been when his mother was walking him to church. Morton carried himself well, exuding an air of self-confidence at every turn. Something always seemed to be happening in his vicinity.

"Enterprising and intelligent," said Sarah Josepha Hale of William Morton. Later a famous editor, she first met him in Boston, where they both worked for a magazine called the *Christian Witness*.

"He never, at any time," Mrs. Hale said, "omitted an opportunity of improvement." So plump with praise, her remark also happened to describe succinctly everything that was wrong with William T. G. Morton, if "improvement" is read with a cynical twist.

In 1836, at the age of seventeen, William Morton moved to Rochester, New York, and took a job as a clerk in a dry goods store. Rochester, incorporated only two years before, was a fast-growing city just coming into its own. Located on the Erie Canal, it was a supply center for the prosperous farming district of western New York. Morton chose well when he decided to move to Rochester. It was a fine setting in which to get rich.

Perhaps because western New York had developed so quickly in 1825–35, without much struggle against the elements, its people were anxious to show their sincere gratitude to the heavens. Or perhaps they were just bored, but new religions and cults sprang forth and jousted with one another in the region, sometimes with violence, in order to stake claims to the souls of the newly prosperous.

On Sunday, April 1, 1838, Morton joined the throng and was "admitted to the communion and family" of the Brick Church, a congregation patterned on Presbyterian teachings. The Brick Church had moved into a swanky new building in 1828, a brick church indeed, and ten years later the congregation was a leading force in Rochester. Another parishioner described Morton's admission to the church as a "profession of conversion," though the need for conversion of a young man fresh from duty on the staff of the *Christian Witness* was either a conceit on the part of the Brick Church or an indication that William Morton would do anything to get into the socially prominent congregation, even convert to his own religion.

As a matter of fact, Morton's new business existence happened to start on the same day as his religious one. While undergoing his conversion, Morton met a native Vermonter named Loren Ames.

"The result of our acquaintance," recalled Ames eleven

William Thomas Green Morton, age twenty-four, from a portrait by William Hudson, Jr., in 1844.
(Wellesley Historical Society)

years later, "was the formation of a co-partnership between us which commenced operation on the first day of April, 1838, for the 'buying and selling of West India Goods,* etc.'" How the partners divided the work at their new store is not known for certain.

Within the first month, however, Ames noted that William Morton had engaged in the following activities: passing bad checks in New York City; adding a phantom five-hundred-dollar entry to the company books ("so that if our neighbors or others looked at the books, they could see that we were doing business upon that amount of capital," was his explanation); surreptitiously changing the name of the company to "Wm. Morton & Co." and then, thinking better of it, renaming the firm after Loren Ames's wealthy brother, Lyman, of Buffalo, entirely without permission.

Four months after Morton & Ames opened, it closed, with a slam that could probably be heard all up and down Main Street in Rochester.

*A general term for exotic imported items, including silk and other fabrics.

Morton couldn't stand still, however. During the prosperous days of his West India dry goods business (that is to say, August), he had realized his cherished dream of taking care of his family. To that end he had placed his sister Elizabeth in a prestigious school called Seward's Academy, in Rochester. Meanwhile Morton remained in the dry goods business, operating under his own name.

Within three months "William T. G. Morton Co." was besieged by creditors, and Morton was facing ruin when he applied for refuge with Phineas B. Cook. Cook was no gentle soul: He was the hardest of businessmen, with extensive properties around the city to show for it. When Morton asked for five thousand dollars' credit, Cook extracted a drastic contract: "I did endorse his paper," Cook said later, "received from him, a judgment bond and levied an exemption upon his goods." The understanding was that the first cash received from the sale of inventory would be used to retire the debts backed by Cook's bond.

Morton kept on selling his goods, and he also kept on keeping the proceeds for himself. When Cook inquired cantankerously into the matter, threatening to sue the young deadbeat, Morton's attorney replied that if his client were taken into court, he would "plead his minority." That defense held that since Morton was under the age of twenty-one when he signed the bond, he didn't owe Cook a nickel. The attorney then doled out some advice, the cruelest possible to a person who is in the right: "He advised me to compromise with Morton in the best manner I could," Cook related.

The Brick Church, for its part in Morton's progress, proved far less willing than even Cook to entertain compromise. The minutes of its general meeting for April 15, 1839, record that, "Brother D. Dickey gave Session to understand and be informed that William T.G. Morton, a member of this church, is accused with being guilty of the sin of profanity and also dishonesty with fellow men. The said William T.G. Morton being present did not deny the charges but admitted them to be substantially true." He promised to mend his ways.

In July, Morton was excommunicated from the Brick Church.

In August, Mr. and Mrs. Seward presented Morton with a bill of $120.17 for his sister's tuition. The reply they received from Morton was datelined Louisville, Kentucky; it assured them that they could draw the amount due from his account at the Bank of Kentucky, sixty-three days from the date of his letter. In the meantime Elizabeth went home to Massachusetts. Soon thereafter—sixty-three days after the date of Morton's letter, to be exact—Mr. Seward realized that "Morton's letter from Louisville was a mere ruse to enable his sister to get away from Rochester without detention of her effects, which were numerous and valuable." Morton's check had been drawn on an account that did not exist, at a bank where he was completely unknown, in a city he'd never been to at all.

William Morton went on to gain widespread fame as a cheat in the 1830s, but it was in Rochester that he first sharpened his tactics. In 1841 the *Rochester Daily Democrat* looked back on Morton's start with what seemed to be clenched teeth: "If he is not as bad as represented," the editors wrote, "all who know him here have entirely mistaken his character." The paper termed any probability of his ever being reclaimed "almost hopeless."

That was Rochester, but then there was Cincinnati, where Morton moved next. It was a fresh field, full of people who wanted to do business with an attractive, personable man who was the nephew of the governor of Massachusetts.

Morton was not the nephew of the governor of Massachusetts, Marcus Morton, but somehow word got around town that he was. There were even those who argued that he was not the governor's nephew but his son. Morton was careful never to give that impression, though: He cast his father as "a heavy dry goods merchant in Milk Street, Boston." The uncle, it seemed, could be famous, but the father should be rich.

By the end of 1839 Morton was in a partnership with Charles S. Pomroy, a well-connected young man with a fine Cincinnati name and even less sense than Loren Ames of Rochester: He left the accounting entirely to Morton.

Morton bought goods on credit (extended against the falsified $6,593.27 he showed on his books), sold the goods, and kept the money. Living quite well on his scheme, he entered into the social life of Cincinnati and with his genuine good looks and supposed pedigree, he made a great hit. He also became engaged to the daughter of one of the richest families in town. That family being Jewish, he was expected to conform to certain expectations, one of which was that he be circumcised, a procedure he had performed by a local doctor. As far as is known, the circumcision was the only incident of surgery and its associated pain in Morton's whole life .

In early 1840, however, the creditors of Pomroy & Co. became anxious and demanded further indications of creditworthiness. "Mr. Morton called upon me on one occasion while I was supposed to be his creditor," recalled a merchant named John Creagh, "and placed in my hands a letter addressed to me, to his care, purporting to be from a firm in Boston (with the Boston post office stamp), well-known as one of the most reputable firms in that city, recommending him in strong terms to my confidence and the confidence of his creditors."

How, one might ask, did Morton arrange to have such a letter sent from Boston so quickly?

Somewhere between Rochester and Cincinnati, Morton acquired a set of U.S. Mail seals, which could be used to print the cancellation marks of several major cities. Morton probably wrote the letter to John Creagh himself (or dictated it to a cohort), put it in an envelope, and then canceled the stamp with the Boston postmark. The forged letter delayed the creditors but not quite enough, and he was eventually caught in his swindle.

Morton was forced to flee without his fiancée. He not only jilted her, he skipped out without paying the doctor who performed the circumcision.

George Bates, a businessman in Cincinnati, was called upon to investigate Morton's business affairs in the city. "He was a young man," Bates reminded himself at a distance of years, when asked to recall William Morton. "The cunning, unscrupulousness and boldness with which he carried on his operations did in my belief mark his career as being without parallel among the swindlers who from time to time have infected our city."

Bates added that "Morton's ignorance of matters of general knowledge was the subject of common remark among all." That impression was echoed by others who knew William Morton well: He just didn't know much. But it didn't matter what they thought in Cincinnati, because Morton had already moved to St. Louis.

"The general expression of his countenance is bold and assuming. He dresses in good clothes, but foppishly. . . . His voice is coarse and loud; he is remarkably ignorant of matters of general knowledge, and is very apt to betray his ignorance by the silliness of his remarks." That is what a St. Louis businessman wrote of Morton after realizing that he was a scoundrel.

This is what the same man also said of him before the realization dawned:

> We take pleasure in introducing to your acquaintance Mr. W.T.G. Morton, late a resident of Cincinnati and now of this place . . . . From our acquaintance with him we esteem him as a young man of good moral and business habits. He is a member of a highly respectable family of tried and close business application . . . . Any favors you may show him, we shall acknowledge with thoughtfulness and cheerfully reciprocate.

The letter was signed for J. B. Sickles & Co. of St. Louis.

James B. Sickles was a dealer in saddlery and other equipment. William Morton and Sickles's bookkeeper had met as roomers in the same boardinghouse, where Morton made himself known by making what a mutual friend called, "ostentatious displays of piety by loud reading and praying so as to be overheard." Morton also let it be known that he made calls of spiritual consolation on prisoners condemned to die. On a higher social plane, he joined the leading Episcopal church and often dined with the pastor, Rev. J. J. Peake. Sometime in 1840 he became engaged to a St. Louis girl and, in the custom of the day, he gave her a miniature painting of himself as a symbol of the engagement.

The glowing letter of introduction from Sickles & Co. was dated August 22, 1840, and with it in hand, Morton hopped on a steamboat going south on the Mississippi River. He was headed for New Orleans, where Sickles's recommendation assisted him in buying a large inventory of goods on full credit. Only days later Sickles was stunned to receive papers from Cincinnati "showing Morton to be an imposter and capable of any act of fraud to which his courage and ability was equal." The result was a memorable panic in the Sickles offices. Letters were sent to New Orleans "by every available conveyance," letters beseeching the merchants there to disregard the firm's previous communication, the one that referred to Morton's good moral and business habits, and to withhold all credit. After that there was nothing to do in St. Louis but wait—and trade stories about how ignorant and coarse William Morton was.

By the time any of Sickles's letters reached New Orleans, Morton was already loading heaps of fine clothing onto boats, allegedly for shipment overseas. Some of the ships had even left port and were headed out to sea. In the very nick of time, the goods were seized and returned to the wholesalers, with no real harm done. Morton returned to St. Louis, apparently intending to rebuild his life there. All was not lost; after all, he still had a rich fiancée. Morton's next goal was to get himself married to her. He steered clear of J. B. Sickles & Co., and took rooms at a different boardinghouse. One day, though, Mr. d'Lange, Sickles's bookkeeper, spotted him in the street.

D'Lange later maintained that William Morton intended to kill him and so he brought a friend—a lawyer named Benjamin Dayton—when he paid a visit to Morton's rooms.

"I had heard that Morton designed to shoot me," d'Lange said. "Upon coming into his room with Mr. Dayton, a pair of pistols lying if I recollect rightly, on the mantelpiece, were taken possession of by Mr. Dayton." On the threat of exposure in the "public prints," d'Lange gave Morton twenty-four hours in which to leave St. Louis.

"Morton seemed greatly downcast upon being threatened with public exposure," d'Lange continued. "He wept and threatened to commit suicide."

However, after d'Lange verified that Morton had not left within a day, as specified, he published a long article in the *St. Louis Evening Gazette* entitled "Beware a Villain." It offered such observations as: "There is no point of villainy at which conscience would induce him to pause. His mind must be constantly occupied in concocting schemes of deliberate rascality, so elaborately planned, so atrocious in their nature . . ." The article predicted that Morton would end up on the gallows.

Morton's fiancée seems to have read the piece. The miniature of Morton's face was delivered to d'Lange the next day, with the request that it be returned to Morton. Yet another engagement was over.

D'Lange may have known all about Rochester, Cincinnati, and St. Louis, but he didn't know about Baltimore. That is because Morton hadn't been there yet. What is more to the point is that Baltimore didn't know about Rochester, Cincinnati, and St. Louis.

After nearly being caught by the Baltimore police with the post office seals in his possession, Morton, so recalled his traveling companion, an acquaintance from Cincinnati, "immediately made his escape from the city, taking with him a new broadcloth cloak belonging to me."

As was the pattern in Morton's travels, the local newspapers then had to make room for an article describing Morton and his villainy in seething detail.

Morton's crimes were not forgotten—if, in the case of the post office seals, they were indeed crimes. Twenty years later, on

March 25, 1861, a front-page editorial in the *Boston Evening Transcript,* having noted the fact that workers were stealing valuable autographs from the papers of the State Department, went on to regret the fact the U.S. laws were so incomplete. "In 1841 or '42," it noted,

> the post office seals of the United States, one of the Boston and the other of the Rochester, N.Y. Post Office, were counterfeited and had evidently been much used . . . yet, as we were informed by a United States attorney, there was no specific law existing at the time the forged documents were discovered, for punishment of the offense.

The case of the counterfeit seals was not forgotten, even at that date, though the name of the perpetrator, William Morton, was withheld. By 1861 that name was world-famous in quite another respect.

Upon leaving Baltimore, Morton made a short detour to Washington, D.C. There, after reneging on a personal loan of thirty dollars, he slipped out of town. At twenty-one Morton still had plenty of time left, but unfortunately he was running out of cities.

Nonetheless, he made his triumphal return to Charlton as a man of means—only because a relative had left him a small sum of money. While William was at home, the family received a visit from a dentist on rounds: Dr. Horace Wells of Hartford, Connecticut. Turning at last toward a respectable profession, William decided to use his inheritance to support himself while he learned dentistry under the auspices of Dr. Wells.

In early 1842, Wells helped Morton to set up his own small practice in Farmington, an elegant old village lined with beautiful houses and populated by a number of substantial New England families. One of the most impressive of the homes was the Whitman mansion, dating from colonial times. The first Whitman in Farmington, who arrived in about 1700, was a minister recently graduated from Harvard College; his father had been in Harvard's first formal graduating class: the class of 1668, which numbered four altogether.

When William Morton arrived in town, the Whitmans' daughter, Elizabeth, was about fifteen, a student at a very prestigious school, Miss Porter's, which is still in operation in Farmington. He decided to marry her the moment he first saw her. He noted that fact in his diary, along with a detailed description of the clothes she wore. And why wouldn't Morton want to marry Elizabeth Whitman at first sight? She was well-educated and exceptionally pretty, with very delicate features, her dark hair in fashionable ringlets. More to the point, Elizabeth Whitman came from a rich family, just as had each of Morton's previous fiancées.

"Dr. Morton paid me attentions, which were not well received by my family," Elizabeth later acknowledged, "he being regarded as a poor young man with an undesirable profession. I thought him very handsome, however, and he was very much in love with me."

By 1843 Morton had risen markedly in his profession, entering into a partnership with Horace Wells for the purpose of marketing Wells's advanced method of making bridgework. According to plan, they opened a Boston office, which Morton was to run.

To the Whitmans, however, a dentist was still a dentist no matter the city. They refused to consent to the marriage until William Morton was studying to become a doctor. According to later accounts, Morton didn't stand much of a chance of actually becoming a doctor, but he noted the loophole, and no one with such poor credentials ever made a better show of *studying* to become a doctor.

Morton had met Dr. Charles Jackson the previous October, when he and his partner, Horace Wells, had asked the famous chemist to analyze the dental plate that they'd invented. Wells returned to Hartford within days, but Morton drew closer and closer to Jackson and his elite circle of friends in Boston's scientific community. He managed something of a coup in March 1844, with his arrangement to start preliminary studies under the tutelage of Dr. Jackson. With Jackson's name floating around in Morton's future, the Whitmans consented to the marriage, and a ceremony was performed May 29, 1844.

"Never shall I forget my sensation as a young bride at sleeping in a room where a tall, gaunt skeleton stood in a big box near the head of the bed," wrote Elizabeth Morton of the first days of her married life with William T. G. Morton. The Mortons spent their first days of married life boarding in the Boston home of Dr. Jackson. The skeleton Morton had procured for his bedroom was something of a prop; he had not yet advanced to the study of anatomy. However, the skeleton might have been symbolic, in a way. Living under Charles Jackson's roof, practicing dentistry as a profession, enjoying marriage to Elizabeth Whitman—just that easily, William Morton left his past behind.

Or at least most of it: Phineas B. Cook caught up with the dentist and, through lawyers, demanded repayment of the money Morton had signed for in 1839 back in Rochester. Morton balked and informed Mr. Cook, through his lawyers, that as he was a minor at the time he signed the bonds, he did not have to repay the money—ever.

Cook's lawyers responded in terms as strong as Morton's. "Whereupon," Phineas Cook later deposed, "his father-in-law wrote . . . that if I would take some two hundred and fifty dollars for the judgment and assign it to him, he would send me the money."

So it was that the Whitmans gained a son-in-law.

William Morton's career as a scoundrel was rarely mentioned publicly during the rest of his life, and it is omitted from most histories to this day.

In his younger days William Morton was a con man. He was an opportunist in romance and a remorseless marauder in affairs of business. Those qualities should not be concealed. If he had not been "bold and assuming," to use the words of one of his early accusers—"impetuous, unremitting, and reckless," in the words of a surgeon at Mass General—then he never would have carried off Ether Day, which was, in William Morton's hands, the greatest confidence game of the whole century.

8

NEXT WHAT?

John Collins Warren and the other surgeons in attendance on Ether Day wanted most of all just to do it again. They wanted another look at the new surgery, with its uncanny quiet and serenity. William Morton was invited back to Mass General the next morning, Saturday, October 17, to give his vapor to another patient, a woman who was having a tumor cut from her upper arm. Dr. Warren assigned Dr. George Hayward to perform the operation. Hayward was a short, stocky man. "His small, deep set, twinkling eyes suggested a certain *bonhomie* and a most amiable disposition," a former student once said of him. "Although not a brilliant operator, his hand was guided by excellent judgment."

The operation Hayward performed in the operating dome at Mass General on Saturday lasted seven minutes: quite a long time, but Morton administered ether at intervals throughout and so the patient did not wake up until after the surgery was finished. Waving aside questions as to whether she had felt any pain, she insisted instead on telling the assembled surgeons and students about the dream she'd had while the tumor was being cut away, about a child she'd left at home.

That operation was such a success that it received top billing in Dr. Warren's diary. The entry for the day of the first operation, October 16, hadn't. It had opened with comments about the mastodon bones, ending with: "Did an interesting operation at the Hospital this morning while the Patient was under the Influence of Dr. Morton's preparation to prevent pain—the substance employed was Sulphuric Ether." But on Saturday, October 17, the priorities had changed. The mastodon was bumped aside: "An operation this morning at the Hospital," Warren wrote in his notational style. "The patient under the influence of Sulphuric Ether inhaled—the patient went through the operation with entire unconsciousness of pain. . . . The experiments on the Mastodon bones are thus far very satisfactory. . . ."

That same Saturday, readers of the *Scientific American* who riffled through their latest issue to page 32 were treated to a short notice of Morton's miraculous new vapor. The October 17 issue of the magazine would have been on the presses before Gilbert Abbott actually underwent his surgery on the sixteenth, and so the report was rather vague, probably based on Morton's use of the vapor on dental patients over the preceding three weeks. It appeared under the headline NEXT WHAT? The article was the first national article regarding etherization.

The full text was as follows:

> Animal magnetism, with all its boasted advantages in rendering people insensible to pain, appears likely to be superseded by a discovery of Dr. Morton, of Boston. It is no other than a gas or vapor, by the inhaling of a small quantity of which the patient becomes immediately unconscious, and insensible to pain: thus giving an opportunity for the most difficult and otherwise painful surgical operations, without inconvenience.

The editors of *Scientific American* didn't know the precise composition of Morton's vapor, but then, neither did anyone else. According to Dr. Hayward, the exact ingredients of Morton's compound represented, if not a true mystery, then a vexing power play. Anyone with a nose knew what the main ingredient was—sulfuric ether has a sickly-sweet smell that no amount of perfume can hide. Officially, however, the composition of the vapor was a secret—a fact that annoyed Dr. Warren very much and Dr. Hayward even more. For his part, though, Morton was otherwise distracted.

In the aftermath of the first two operations, and the local publicity inspired by them, William Morton busied himself in his dental practice, trying to accommodate a crush of patients. He pressed his advantage with a barrage of advertisements placed in the newspapers for his unique service: "Teeth Extracted Without Pain." He later claimed to have administered ether to two hundred patients during the month of October, the first full month in which he employed the vapor in his practice. Assuming that he didn't work on Sundays, and recognizing that he lost at least three days demonstrating the compound at Mass General, that averages nine or ten etherizations per day. The fact, though, is that William Morton administered few, if any, of them; he left the etherizations to his assistants. The word among the office helpers was that the boss wasn't good at gauging the use of ether, and didn't much like to do it. Morton experimented very little outside the operating theater, a fact that made his success there all the more remarkable.

However, Morton's assistants, having no medical training whatsoever, were also having uncertain results. Ether may have been a boon for Morton, but unfortunately it was also a boon for Dr. Augustus Gould, his friend and landlord. Whenever ether made a dental patient sick, Morton or one of his assistants rushed to ask Dr. Gould's advice. In six or eight cases Gould recalled attending Morton's patients personally. "One was in a state of very high excitement, almost a maniac; she was brought to my house," Dr. Gould said. "Some others vomited profusely. One or two were lethargic, and roused with great difficulty." It is plain lucky that Morton's demonstrations at Mass. General were carried through so faultlessly. Had either of his initial patients there become hysterical, vomited profusely, or remained lethargic to the point of alarm, Dr. Warren might well have stepped forward and told the gentlemen in the dome that Morton's compound was indeed a humbug, and a colossal one.

In hindsight people who knew William Morton reckoned that he would have been happier had he shirked the opportunities surrounding hospital surgery and concentrated all his efforts on using etherization only to expand his dental practice. He understood the world and the way it worked well enough to do that—but not much more.

Almost immediately he started organizing a national sales force to represent his compound. On October 19 he even wrote to Horace Wells, thinking he could offer his impecunious friend a job selling the compound. Without quite knowing what Morton had discovered, Wells wrote back enthusiastically, saying that he would be in Boston within a few days to meet with him.

Charles Jackson was still growing bilious at any suggestion that he should attach his name to the discovery, since William Morton, whom he considered a vulgar man, was linked with it in newspaper articles and advertisements in Boston. However, Dr. Jackson's attitude would change when he heard Dr. John Warren's opinion of the new discovery.

Dr. Warren spent Tuesday, October 20, overseeing the transport of his mastodon skeleton from Harvard to his home. He also spent much of the early part of that week planning the first meeting of the Thursday Evening Club.

Dr. Gould received an invitation to the meeting, as did six other selected guests. Abbott Lawrence, for example, was a merchant and cofounder of Lawrence, Massachusetts. He had donated fifty thousand dollars to support the study of science at Harvard. Theophilus Parsons was a law professor at Harvard, living comfortably in the shadow of his father, also named Theophilus, a forceful Massachusetts politician in the first days of the Republic. Charles T. Jackson was there too: A graduate of Harvard Medical School, he had devoted his whole career to the sciences. He was very well acquainted with J. C. Warren, a generation his senior. For his part Dr. Warren admired Jackson and once said that he was "distinguished for his philosophical spirit of inquiry, as well as for his geological and chemical science."

The new club's first meeting must have opened with a look at the mastodon bones, though it is doubtful that they were erected. During the general conversation that followed, etherization was naturally the subject of immediate interest, especially among the medical men. Those who had not heard about the discovery were fascinated, as a tremendous future was predicted for painless surgery. After the discussion had continued for a few minutes, Jackson piped up to claim that it was he who had given William Morton the idea for etherization. That declaration came as a total surprise to Dr. Warren.

The very next day Jackson went to see William Morton and surprised him, too, demanding a steep fee of five hundred dollars against 10 percent of the revenues from the compound, for the advice rendered on September 30. Even so, Jackson was still adamant that he wanted no public acknowledgment of those services. He made his demands on October 23. That evening Morton's lawyer, R. H. Eddy, arrived at Jackson's office. Eddy talked Jackson into something else entirely: a joint application for a U.S. patent with William Morton.

Privately Eddy's maneuvering was generated by his belief that the patent for etherization would not hold up unless both men were named on it. To sell Jackson on the idea of applying jointly, Eddy suggested that should Jackson "not do so, he might lose all credit, as in the case of the Magnetic Telegraph, which I had understood from Dr. Jackson he had suggested to Professor Morse." It was a canny remark, one worthy of the lawyer. Up to that time the bitterest experience in Jackson's whole life had been losing what he considered rightful credit for the telegraph to Samuel Morse. It was the mention of the telegraph that turned Jackson around, to his later regret. The best that can be said of it is that thanks to Eddy's advice regarding the discovery of ether, the telegraph episode would no longer be the most painful in Jackson's life.

Dr. Jackson agreed to apply for the patent with Morton, and to split the profits: Morton was to receive 65 percent of revenues; Eddy, 25; and Jackson, 10. Dr. Jackson was satisfied with the promise of some thirty or forty thousand dollars, and he seemed to cooperate with the cocredit idea for the next week or two. During that time Eddy prepared the application for a U.S. patent, which would be formally entered on October 27. Jackson may have been placated. Horace Wells, however, was not.

Wells arrived in Boston on October 24, the day after Jackson agreed to participate with Morton in the patent. He had made a special trip from Hartford with his wife to see Morton and the new compound. According to the arrangements made in advance, Wells observed the use of ether in William Morton's office. After his visit he was thoroughly disgusted by what he'd seen. Morton later recalled that his former friend and partner "said I should kill someone yet and break myself up in business." He also remembered that Wells left the office very abruptly.

"It is my old discovery," Wells told his wife as soon as they were outside. "And he does not know how to use it." In disgust the couple left Boston for Hartford, where Wells renewed his efforts in yet another business venture, yet another unfortunate one, as it turned out, intended to market a coal sifter he'd invented.

For his part Morton didn't care about Wells or anyone else. Busying himself conspicuously with his own dental practice, he was even refusing Mass General's requests that he detail the composition of the vapor.

"Ether was evident, but he pretended that there was some other ingredient," wrote Mass General's admitting physician, William H. Thayer, concerning Morton's attitude. Knowing perfectly well that sulfuric ether was the active ingredient, and having an abundant supply of it at the hospital, J. C. Warren directed Dr. Hayward to find a glass globe and hook a tube up to it, in the same manner as Morton's apparatus. He was looking forward to attempting etherization during the hospital's usual schedule of surgery at the end of the week, on October 30 and 31. At that point, however, someone had to tell John Collins Warren about the patent, and that the use of ether without Morton's permission would constitute infringement. It was not an opportunity to vie for.

Having built his career—and, not incidentally, Boston's foremost hospital—by honoring medical ethics and respecting political expediencies, J. C. Warren at sixty-eight was not a man to allow his intentions to be checked. . . .

"I was checked," he later recalled, "by the information that an exclusive patent had been taken out, and that no application could be made without the permission of the proprietor. The knowledge of this patent decided me not to use, nor encourage the use of, the inhalation."

That was the position forced on a man who told every medical class he ever faced that surgeons throughout history had been in dire need of a painkiller for their patients. During Friday's operations and then during Saturday's, Dr. Warren, Dr. Hayward, and the other surgeons on staff at Mass General caused sufferings that they were not, as a matter of fact, obliged to inflict. Dr. Warren had no desire to squelch the discovery of etherization permanently; he recognized its potential to revolutionize surgery. However, confronted with a patent on a medical breakthrough, John Collins Warren faced a revolution of a different sort. If October 16, 1846, was to be a dividing line in humanity's lot, then October 27—the formal date of the patent application—heralded another type of new era, one in which business considerations encroached on the hallowed ground of medical advancement, compromising its spirit but also accelerating its pace.

The first half of the nineteenth century was enthusiastically hooked on pills, as well as tinctures, powders, drops, syrups, and any other form of remedy that could be pushed across a druggist's counter. In fact, the establishment of trade-name medicines in the early 1800s has actually been attributed to the weak state of surgery: In those many cases too daunting for either the surgeon or the patient or both, some elixir proffered in some store somewhere offered hope, at least. At the time, any remedy sold over the counter in a store (or on a street corner) was considered by physicians to be a "patent" medicine. Many of the remedies were, indeed, patented, but the point was that all of them were moneymaking ventures.

To legitimate practitioners "patent" medicines were inherently unethical: If a compound was of any use in the relief of disease or suffering, it had to be made freely available to all.

Physicians were not great moneymakers in the nineteenth century. They were fortunate to make a middle-class living; only a small percentage actually accumulated anything like a fortune through their practices. Many physicians worked hard just to keep themselves from slipping even further down the scale financially. Some concentrated on it too much: When Edward Jenner's discovery of vaccination reached the United States in about 1800, doctors from the top of the profession on down rejected it, for many of them did not want to *eliminate* smallpox: They wanted to treat it. Patients, however, were not so sentimentally attached to the disease, and they soon succeeded in breaking whatever clusters of resistance had formed at first.

Many of those who were outraged over the ether patent would later point to Jenner's 1798 discovery as setting a suitable precedent in a different respect. Jenner, an Englishman, hadn't tried to patent his discovery but rather had worked for years to propagate it. Then, after it was universally accepted, he received a generous pension from the British Crown, along with cash awards from at least six other governments. The covenant kept with medical researchers throughout Europe was that if they were to give a gift to medical science, they could expect to receive a gift in return, from the government of a grateful people. The same arrangement had never been tested in the United States—since no lofty discovery had been made there through 1846. William Morton, who was part of neither a European tradition nor a medical one, shattered the covenant when he conceived the idea of taking out a U.S. patent on his decidedly useful discovery. He would make sure of everything in advance, and he would extract his reward.

But not from Massachusetts General Hospital.

Dr. Warren picked his way through the budding controversy and decided that such a magnificent discovery was not going to be controlled by one man, William T. G. Morton. At any rate Dr. Warren wasn't going to be controlled by Morton.

On October 28 he apparently discussed the matter in person with Morton, for he wrote a note to him on that date: "I had the pleasure to call on you today to converse on the subject of the Gas," it began:

> I am very anxious to find a mode of mitigating the sufferings of patients under surgical operations. If you can, without impropriety, give me a practical account of the apparatus, and the substance employed; or purchase for the Hospital this apparatus, it would be a real blessing to humanity, and a favor to . . . Your friend and servant, J.C. Warren.

Morton declined to reveal the information Warren requested. His recalcitrance led to an impasse that lasted several weeks. On November 1, according to schedule, Dr. Hayward took over from John Warren as chief surgeon at Mass. General. He intended to take a hard line, so he told Dr. Warren the next day.

During the delay over the controversy, Mass General's surgical staff began to divide. The secrecy surrounding the discovery and the patent protecting it represented a future that senior surgeons could not reconcile with the open spirit of the past. Henry Bigelow, the youngest member of the surgical staff at twenty-eight, felt no such tug from the past and to him the delay—any delay—was damning to them all.

The son of one of Boston's best-known physicians, Henry Bigelow had followed his father's path through Harvard and its medical school. The sort of boy who had been interested in practically everything around him, Henry was exactly the same type of man. As an undergraduate he participated in a college rebellion so rocky that it brought parents from all over to the campus in Cambridge. Dr. Bigelow tried to persuade his son, Henry, to quit the rebellion, but the lad refused, finally reminding his father that there had been a rebellion back in his day too. Dr. Bigelow did not deny it: "Yes," the father admitted. "But I have seen the folly of it."

"Well, I want to see the folly of it too," Henry replied. After taking a turn in Paris, and then his medical degree from Harvard, he opened a surgical dispensary: A rather radical idea for its day, it was akin to an outpatient clinic. Through his own attainments and the influence of his family name, he was appointed to the surgical staff of Mass. General in 1846. Alone among his colleagues, he used late October's unfortunate hiatus in etherizations to continue observing the Morton technique. For him it was the very moment of discovery that was fascinating. To miss it would be akin to being alive during the passing of a great comet and not bothering to look.

During the same year, 1846, Bigelow gave a speech refuting the "numerical" method of medical inquiry, which held that imagination had no place in research: If physicians kept detailed records of cases and treatments, the theory ran, then the plain truth was bound to emerge from a broad enough field of statistics. Bigelow contended that the numerical theory eliminated the spark of genius and that statistics were best used to support a hypothesis generated beforehand by the human imagination. Statistics could not replace that imagination, not having its commanding view. In the essay Bigelow then supported his own hypothesis with facts from the history of medicine. His most recent example—Edward Jenner and the development of vaccination—was already about fifty years old. He'd read about that in books. But with etherization Henry Bigelow had an invention on a par with any, and he had it right before him.

So what if it was patented?

Bigelow plunged ahead, and if there was any folly in it . . . he wanted to see it for himself. During the first few days of November, he was writing an essay about what he'd seen, particularly at Morton's office, and the importance of etherization for medicine. "On the evening of November 2, 1846," Bigelow's friend Oliver Wendell Holmes remembered:

> he called at my house in Charles Street with a paper which he proposed reading at the meeting of the American Academy of Arts and Sciences, to be held the next day, and which he wished me to hear. . . . He was in a state of excitement as he spoke of the great discovery that the gravest operations could be performed without the patient's knowing anything about it until it was all over. In a fortnight, the news of this wonderful discovery, he said, will be all over Europe.

The next evening Bigelow read his report to the academy, but he had to refrain from publishing it until the dispute with Morton was resolved.

During the same week William Morton received a dismaying piece of advice from his lawyer: It would be difficult to promote the ether compound for surgery unless it was proved in a true capital operation. The operations of mid-October had been relatively simple, the removal of a surface tumor in each case. Morton needed to test it on something more severe. To that end he returned to Mass. General on November 6. Paying a visit to Dr. George Hayward, Dr. Warren's replacement as head of surgery, Morton requested permission to administer the compound in a case of amputation at the hospital. Dr. Hayward was waiting with his answer, and it was conveyed with no sign of that amiable twinkle in his eye.

"I did not intend to allow the surgical patients to inhale this preparation of Dr. Morton during my period of service, unless all the surgeons of the Hospital were told what it was," was Hayward's position, as he later recorded it.

There was no response from William Morton.

9

POWER STRUGGLE

September 1846 had been the month in which the discovery of etherization emerged, when it was taken out of the world of entertainment, after being so long wasted on laughter. October had been the unveiling, when etherization took its place as a medical wish fulfilled. After being introduced at Mass General, it was in the hands of men who would not let it be forgotten ever again: Lions such as John Collins Warren wouldn't let it out of their sight; hungry cubs such as Henry J. Bigelow wouldn't let it out of their clutches.

November was the month of opportunity. It was the month for doubters and interlopers, for sides to be taken and stakes to be raised, for superlatives and absurdities, and for a daily swirl of events stirred by greed of all stripes, the best kind and the worst. November was also the last month in which the discovery remained at home in Boston, for it was the month that sent the discovery around the world.

William Morton could see that November was going to move fast, and that it was taking the ether discovery away, with or without him. On November 6 he sat down to write a letter informing the Mass General surgeons that the only important ingredient in his compound was indeed sulfuric ether. The timing was no doubt carefully considered: The patent application had been made almost two weeks earlier, and the patent was due to be granted any day. Dr. Jackson was out of town at the time, but Morton and his lawyer, R. H. Eddy, were waiting intently for the official issue. The priority of the application made Morton less concerned about revealing his secrets; at any rate no one could steal the idea and jump ahead of him in the patent process. And so Morton wrote his letter, had it delivered to Dr. Hayward on the afternoon of the sixth, and then waited. He knew that having no other affiliation, he needed to ally himself with Mass General. Anyway, he knew that he could not fight it, and the rest of the world, at the same time.

The outcome of the negotiations between Morton and the hospital mattered most of all to an Irish girl of twenty named Alice Mohan. She was the one scheduled to have her leg cut off the next day, November 7. A bone disease had destroyed her knee joint and rendered her lower leg useless, a condition that had kept her in the hospital for a year and a half. Because the disease seemed to be spreading to the rest of her body, the surgeons scheduled an amputation.

It was to be theater, and it was to be good theater. By early November, Harvard Medical School was in full session: Mass General was bustling, and everyone around the hospital wanted to see an etherization. The operating dome was packed to the rafters on the seventh, a Saturday, and spectators still kept crowding in; if they could not find seats, they stood around the gallery, where their heads nearly brushed the ceiling.

Augustus Burbank, a twenty-three-year-old medical student from Maine, wrote a letter to his sister about the scheduled operation. "At 11 am on Saturday, we assembled in the operating theater in the Hospital to witness Surgical operations," he recounted a few days later "Who should take his seat at my right side but Dr. Hutchens of Poland [Maine]. He had come to consult Dr. Warren about his own case. . . ." Burbank then told his sister all about Dr. Hutchens's recent operation, the original account of which contributed its own note to the nervous chatter that filled the gallery before the Mohan operation began.

As of ten o'clock and then ten-fifteen, even as observers

waited in anticipation of a major event, Morton was still at home. As of ten-thirty he was still there. Unlike the drama surrounding the first demonstration on October 16, though, Morton's absence was anything but his own doing. The fact was that the surgeons had yet to decide whether Morton was to be allowed in the building. At eleven o'clock, though, Alice Mohan was scheduled to undergo her surgery, the old way or the new.

"About a half hour beforehand," William Morton wrote, "Dr. H.J. Bigelow [young Henry Bigelow] called for me, and said he wished me to be on the spot, in case it should be determined to admit me." So Morton trooped along to Mass General, at the mercy of a decision yet to be made.

A few minutes before eleven, six surgeons entered the operating dome from one of the side entrances: Dr. Warren and Dr. Hayward were first, followed by the rotund S. D. Townsend, crisply attired and sporting conspicuous sideburns. They were the old guard, but three young surgeons were also on the staff. Dr. J. Mason Warren was J. C. Warren's youngest son; he was an elegant young man and an excellent surgeon—having had remarkably little choice in either the matter of being a surgeon or in being excellent at it. Henry Bigelow, also in his twenties, was in the group: He was a strapping man with sharply drawn features and a rakishly unkempt appearance. The last of the surgeons in attendance was Samuel Parkman; tall and dignified at the age of about thirty, he was keenly interested in the process of etherization and would use it on his own, starting in late November.

As the eminent surgeons made their procession into the room, the audience was oblivious to the fact that Morton was as yet uninvited. According to Daniel Slade, those in the audience were gazing around at the instruments, the skeleton case, and at the mummy that gazed back at them from the walls of the operating theater: "The assembly was scanning these various objects," he recalled, "gazing upward also at the elegant and well-lighted dome and taking cognizance of every trivial incident, as is customary with impatient crowds." Their impatience was of no import to Dr. Warren and the other surgeons, who gathered in a circle and conducted an intent conversation as the spectators looked on.

Dr. Bigelow acted as the emissary between the surgeons and Dr. Morton, who was by then waiting elsewhere in the hospital. After two such conferences, one decision was finally made, at least. The patient was sent for.

Daniel Slade later recalled the sight of Miss Mohan being carried into the room on a stretcher, "the bright hectic flush upon her cheeks contrasting strongly with the white sheet which otherwise enveloped her entire form."

Finally, William Morton agreed through the emissary of Henry Bigelow to allow Mass. General full and free use of his patent, with no rights reserved, no charges made, and no further secrets. Then he, too, was allowed back into the circle of surgeons. Bigelow escorted him into the operating theater.

Slade described his first sight of William Morton: "a man of commanding figure and appearance, very erect, and dressed, as he usually was, in a stylish fashion peculiar to himself, consisting of a blue frock coat with brass buttons, a large and elegant scarf which completely filled up the open front of the waistcoat, 'gaiter' trousers,* etc." Without acknowledging the surgeons, Morton set up his equipment and then spoke gently to Alice Mohan. When both of them were ready, he held the mouthpiece against her lips and began to adjust the valves on the retort.

"The stillness was oppressive, broken only by the hurried respiration and occasional sob of the patient," Slade recalled. He then repainted the scene as he recalled it, including that subtle strain of irony, or snobbery, that would face William Morton whenever he placed himself among great medical figures. "Grouped about Morton," Slade wrote, "standing as the central figure at the head of the operating-table, were the surgical and medical officers of the institution, as also the attendants, all as

---

*Short-legged to show off one's shoes.

intent upon the unusual scene before them as were the most untried spectators in the seats of the amphitheater."

After three minutes Alice Mohan was insensible and seemed to be asleep. In an instant attendants slid her down to the end of the operating table and held her leg straight out from the end. Dr. Mason Warren closed off the femoral artery with pressure from his fingers. At the same time, Dr. Hayward pushed the flesh up from the back of the thigh and then swiped through it with a surgical knife, the angle of his cut forming a flap in the skin. An assistant peeled the flap back, so that Dr. Hayward could make a completing cut through the flesh left on the underside of the thigh bone. While the assistant held both flaps back, Hayward sawed the bone but left a jagged chip on the end, which he clipped off with a pair of forceps. Having sewn the femoral artery and four other blood vessels, Hayward was about to tie a sixth one and then close the stump when the patient stirred for the first time. She was waking up just as he finished. Her only protest was that the surgeons hadn't yet taken her leg off—that certainly impressed everyone who was watching. One of the attendants held her dismembered leg where she could see it.

According to Dr. Hayward, Alice Mohan was the first person whose life was actually saved through the use of anesthetics. In his opinion she could not have withstood the operation otherwise, and "would have sunk," in the euphemism of the day, meaning that she would have died.

Following Miss Mohan's operation, Dr. Warren performed a far-more-protracted operation in which he removed part of the jaw of a middle-aged man, painlessly, due to the influence of ether.

"Morton was the hero of the hour," Slade wrote, "and was regarded with feelings akin to those which might have been awakened had an angel suddenly appeared."

If Morton appeared as an angel, he appeared as a different type of angel to everyone in attendance. To a student hungry for practical knowledge, such as Augustus Burbank of Maine,

William Morton did not bear a gift merely for humanity but one for Harvard medical students. "A great deal of excitement prevails," Burbank wrote to his sister, after giving her a description of the operation,

> and well there may. As the surgeons of the Hospital sanction the use of this gass [*sic*], we shall probably have a plenty of operations at the Hospital this winter. In all probability, many will resort thither, for Mr. Morton will protect his discovery with a patent and therefore those who desire a painless surgical operation will be obliged to come to the Hospital, the theater of its usefulness.

Augustus Burbank would not be the only person in the medical profession to perceive the fact that Morton's reviled patent seemed to have become ethically quite acceptable to Mass General, just as soon as the hospital was let in on it.

On November 12 the patent was officially granted, through the normal channel of the day, by the secretary of state (and future president), James Buchanan. On that same day Dr. Warren expressed sheer delight about a prank played on him by ether. He was standing right next to a patient about to have a tumor removed from her arm, under ether, and he missed the whole operation. "So entirely tranquil was she, that I was not aware the operation had begun, until it was nearly completed," he marveled.

All over Boston people were trying to describe the angel, and just like Augustus Burbank, to put themselves somewhere in the depiction.

By coincidence Harvard University was set to dedicate its new medical school buildings, near the hospital, in ceremonies scheduled for November 12. Edward Everett was scheduled to deliver the keynote address. Everett, the president of Harvard University at the time, is best remembered for being so much forgotten: In November 1863 he was the keynote speaker at the dedication of the national cemetery at Gettysburg, Pennsylvania, though it was Abraham Lincoln, following him rather humbly, who delivered the great Gettysburg Address. At Harvard's medical school in 1846, however, Everett read his prepared speech and then included a hastily added paragraph at the very end, citing the discovery of etherization as having been introduced at Massachusetts General Hospital—and under the auspices of Harvard professors.

However Edward Everett may have felt about that day in Gettysburg, he was proud until the end of his days to claim a measure of credit regarding the discovery of etherization, often citing himself as the first person to mention it in public, outside of a hospital.

Morton, meanwhile, was making his own claims, where it mattered most to him, placing ever bolder advertisements in Boston newspapers for his painless dentistry. The *Boston Daily Evening Transcript* of November 20 contained two of his ads, both of them on page 2. "Teeth Extracted Without Pain," promised the one in column 3, inviting patients to visit the office to have a tooth extracted or to examine certificates affirming the attributes of the compound.

The ad in column 2 was more ominous. In it Dr. Morton gave "public notice that Letters Patent [had] been granted by the government of the United States on his improvement," and that while interested dentists and surgeons could inquire about buying a license to use the patented process, "all persons are hereby cautioned against making infringements on the same."

That ad was apparently a waste of money.

"Extracting Teeth without Pain," began an ad over in column 5, "Ross & Houpt are prepared to extract teeth by their new improved compound, which has the advantage of all ethers now in use . . . . Price of extraction reduced to $1." The discovery was only weeks old, but it was already on sale!

While Ross & Houpt had commandeered William Morton's discovery, most other Boston dentists were renouncing it. In fact many of Morton's brother dentists had banded together just for that purpose, led by J. F. Flagg, scion of a family long associated with progressive dentistry in Boston. During November, Flagg instigated an articulate debate about the ethics of taking out a patent on a "gift to humanity" and about the legality of awarding protection to Morton on a compound as common as sulfuric ether. A few dentists, including Ross & Houpt, saw no reason to debate any of the fine points—they simply used ether and waited for someone to stop them. No one did. Morton soon realized that he would have a better chance of stopping Niagara Falls than of prohibiting medical practitioners from administering sulfuric ether. However, he was thinking the modern way when he realized that every invention needs a little *invention* to make the public see it the right way and want it. And so Morton called a meeting of his inner circle of friends on the night of November 20.

The meeting was led by Dr. Augustus Gould, in whose living room it was held. The others in attendance were Dr. Oliver Wendell Holmes, Dr. Henry J. Bigelow, and Professor Louis Agassiz,* the acclaimed geologist from Switzerland. Holmes, who was new to Morton's circle, filled an unusual role in Boston's intellectual community. He had been trained as a physician, and earned respect early in his career for an insightful essay on puerperal (postpregnancy) fever. Yet the many medical societies that invited Holmes to speak discouraged him from trying to be intelligent: They only wanted him to be amusing. And he was, effortlessly. Magazine editors besieged Dr. Holmes for contributions, but they weren't after a continuation of his work on puerperal fever, they only wanted his lighthearted essays and verse. (Holmes's son, Oliver junior, a five-year-old in 1846, would serve on the U.S. Supreme Court for thirty years, starting in 1902.)

At a time when Boston was bulging with powerful intellects and very attractive personalities, William Morton's brain trust

---

*Pronounced *AG-ga-see.*

included four of the most impressive men in both respects. Their immediate mission was to conjure a name for Dr. Morton's compound, one that would give it a cachet beyond the overly descriptive "etherization."

"Etherization" was practically a recipe, all by itself. The new name would have to imply something more—a process, a singularly Mortonian process that would attract customers—that is, patients—to it even in preference to plain ether. This was far more powerful than any mere patent. The goal was a trade name. It was not law, it was business.

And it was not medicine, at least not the way the old guard thought it should be. If it is hard to think why Boston's establishment should assist the likes of William T. G. Morton in fighting the Ross & Houpts of the world, Oliver Wendell Holmes supplied a hint, in the letter he wrote to Morton the morning following the brain-trust meeting. As it turned out, Holmes didn't like the name that Morton had rather temperamentally insisted the others adopt: "Letheon."

"Everybody wants to have a hand in a great discovery," he wrote to Morton, with his disarming frankness. "All I will do is give you a hint or two, as to names . . . ." His choice for the insensible state was "anæsthesia." The adjective that he suggested was "anæsthetic." The words emanated from the Greek for "insensible"; it had been in use before to denote parts of the body benumbed but not paralyzed. For example, an 1815 article in a British medical journal called *Medico-Chirurgical Transaction* described "A Case of Anæsthesia," in which a Scottish emigrant to Jamaica had no sense of feeling in either of his arms, though he could move them perfectly. That was anesthesia in the old sense. Holmes actually only borrowed the word for the new state of being, though he has received credit ever since for coining it.

Holmes was certainly right about one thing, though: Everybody wanted to have a hand in the great discovery. That was the mood of Boston in mid-November.

Charles Jackson returned from a geological journey of about

ten days in mid-November, to find that the stalemate at Mass General had been broken and that the operation there on the seventh had been the most convincing demonstration yet. On November 15 the lawyer, R. H. Eddy, called on Dr. Jackson on a routine matter related to the patent (granted in Washington three days earlier). When Eddy was through, Jackson spoke up. As though nothing had happened before, he informed Eddy that the entire discovery was his. William Morton had played no important part in it whatsoever.

Eddy was "astonished beyond measure," as he put it. A few minutes later, however, he was even more astonished to hear himself agreeing to give Jackson an increased percentage from the overseas revenues to be derived from the patent. Unless Jackson received the increase, he was threatening to send a detailed claim to the Academy of Sciences of France, recognized internationally as the arbiter on such issues. Since the next transatlantic steamship was due to leave Boston the following day, Jackson's threat carried a glaring imperative. Eddy acquiesced and signed over the increased percentages Jackson demanded. And then he went back to his office and prepared a letter on Morton's behalf for the Academy of Sciences. Morton and Jackson were no longer on the same side, if they ever had been.

On November 9 Henry Bigelow had read a new version of his history of etherization, the month-old discovery, to the Boston Society of Medical Improvement; more important, he was encouraged by his elders to publish the work. On Wednesday, November 18, 1846, the lead article in the *Boston Medical and Surgical Journal* was Henry Bigelow's "Insensibility During Surgical Operations Produced by Inhalation." As a platform for the truly amazing announcement, the article did more than anything else to give the new field of anesthetics a life of its own. Young Henry Bigelow rendered the world a great favor by writing such a strong endorsement under the banner of a respected medical journal, but he also did William Morton a great favor by stopping well short of a confirmation that Morton's compound, his Letheon, was just sulfuric ether. In fact, though he knew perfectly well that the compound consisted of little else, he even went so far as to imply that ether alone could not produce the effects of Letheon.

That ambiguity infuriated Dr. Charles Jackson. He rigidly noted that Bigelow's essay concealed the exact nature of the compound: "Hence," he declared, "it is a mere quack advertisement."

As though Henry Bigelow's articulate essay *were* a mere quack advertisement, William Morton took advantage of it as a head start, and formally assigned territories all over the country to businessmen ready to distribute Letheon to physicians and dentists. Letheon, the first boon to humanity to have a catchy trade name, was sent forth at the end of 1846 as though it were a newfangled corset or a plumbing appliance. Many doctors and dentists signed up and started doling out Letheon. A few who were perplexed by the dawning of commercialized medicine did what perplexed people usually do: nothing at all. A swelling number were more angry than impressed, however. The veritable Letheon bazaar that the United States became in 1846–47 caused many medical men to reject the idea entirely, because of the moneymaking framework in which it arrived. That was November.

10

A WORLD WAITING

Though countless Americans were talking about Letheon, very few were giving it a try in surgical operations.

In the United States, the initial debate over the etherization discovery is easy to perceive as a regional one, as members of the medical communities of New York and Baltimore and, even more rancorously, New Orleans and Philadelphia, attacked their colleagues in the self-proclaimed "hub of the universe," Boston. Nothing annoyed other cities more than those instances in which little Boston actually *was* the hub of the universe.

Boston didn't particularly care about the rest of the country, however. In the first rush of excitement all eyes there were focused on Europe, or at least they were focused on the sailing timetables for ships leaving Boston during the month of November 1846. Europe mattered most of all to Americans educated there. For some the new discovery represented an opportunity to please the old master. Others, however—and *one* other in particular—looked to Europe with eyes not quite so wide, knowing that it was the medical community of Paris that arbitrated questions of priority in disputed discoveries. So it was that Charles T. Jackson was among the very first to send a letter overseas regarding the use of ether in surgery. R. H. Eddy was only slightly behind, writing on Morton's behalf.

A handful of Boston's medical men simply wanted to promote etherization in those capitals best at relaying new ideas to the rest of the globe.

Dr. Jacob Bigelow, Henry's father, put a letter to London on one of the fastest ships in the transatlantic service, the Cunard steamer RMS *Acadia,* on December 1, and included in it a copy of his son's article in the *Boston Medical & Surgical Journal.* The envelope was addressed to an old friend, Dr. Francis Boott, a rather insouciant American who had settled in England. Dr. Boott had largely shirked his own practice in favor of a convivial existence in the midst of London's medical circles. In his letter the elder Dr. Bigelow came right out and stated the great American secret, that "the process consists of the inhalation of the vapour of ether to the point of intoxication." Boott received the news from Boston on about December 16 and hurried a copy of Jacob Bigelow's letter to the *London Medical Gazette,* which devoted a long column to it under a headline less than eye catching: "Discovery of a New Hypnopoietic."

All those on alert for hypnopoietics, and those fewer who actually knew what the word meant,* read of the efficacy of ether and the ease of its use in rendering patients insensible of pain. "The facts are here so candidly stated that any one may put the new process to the test of experiment," the *Gazette* stated, adding a sentence that probably explained why etherization would be more quickly embraced in England than in the United States: "Dr. Morton has made no mystery of his proceedings, like the tribe of hypnotic quacks who have lately perambulated the country [England]." Of course the *Gazette* was perfectly wrong: At that very moment Dr. Morton was making an utter mystery of his compound, but thanks to the Bigelows, etherization arrived in England without any of its complications attached. London was too far away to see the small tribe of claimants perambulating all over Boston, and that is just as well.

On Friday, December 18, Dr. Boott gave the letter and the

---

*Sleep inducers.

article regarding ether to Dr. Robert Liston, a superb London surgeon. The next day Liston bought sulfuric ether and tried it out on a young woman about to have a tooth removed; the day after that, he fashioned an inhaler, and on Monday, the twenty-first, he used ether at North London Hospital (now University College Hospital) in an amputation of the leg, with resounding success. According to one of those who was present, Dr. Liston turned to the audience after the miraculously calm and silent operation. His voice carried the same raw sense of drama with which Dr. Warren had addressed the audience at Mass General the previous October 16, when Warren's usual eloquence was reduced to blurting, "Gentlemen, this is no humbug!"

Dr. Liston's eternal words were, "This Yankee dodge, Gentlemen, beats mesmerism hollow!"*

Right after the operation Dr. Liston wrote to Francis Boott, reporting that the ether had been used " with the most perfect and satisfactory results." He then added "I made up my mind to make the experiment only a few minutes before the operation." That may or may not have been true, considering that he did devote most of the night before it to writing out invitations, so as to ensure a packed house at the operating theater. Dr. Liston wrote his note to Boott in a rush, since he was hosting a dinner party at his house that very evening. It must have been a lively party—the host certainly had something to talk about. He also had some ether left over and used it on one of his assistants after dinner, in a further effort to entertain his guests.

The *Lancet*, Britain's preeminent medical journal, printed the correspondence variously written by the Bigelows, Boott, and Liston in its issue of January 2, 1847. During the remainder of the year, the journal would print two hundred letters and accounts regarding the new art of anesthesia.

Dr. Liston's operation launched the practice of anesthesiology outside the United States, but it was not the very first

*A "dodge" was a clever plan.

demonstration. As it turned out, the ship's surgeon onboard the *Acadia* was an alert young doctor named William Fraser. He hopped off the ship in Liverpool and, it seems, jumped on a coastal steamer headed for his hometown of Dumfries, on the southwestern coast of Scotland. In any case he got himself to Dumfries before the Royal Mail got Jacob Bigelow's letter to London. Just like his famous colleague, Robert Liston, William Fraser wasted no time in trying the "Yankee dodge," and he administered it to a patient on December 19. As the operation was never reported to the national medical press, it had no effect outside the Dumfries and Galloway Royal Infirmary, but nonetheless it indicates the gallop at which etherization entered Great Britain.

Within the world of surgery, Edinburgh, Scotland, was recognized as a capital of the art, on a par with London or any other city. On December 23 the two ranking professors of surgery there received letters from Robert Liston, written in his own hand, headed "Hurrah!" The letters, which were substantially alike, went on to describe etherization in ebullient, even zesty, terms. The process was tried in Edinburgh within the week.

One late December afternoon in Glasgow, in western Scotland, a medical school class was kept waiting for a scheduled anatomy lecture, and the students were not patient about it, chanting and throwing things around the room. Finally the professor appeared, grave with emotion, to tell them that the class was canceled. A ship had arrived from Boston that morning with news—the greatest news that had ever been received in surgical science—he said. A painless operation had been performed at Massachusetts General Hospital and an attempt would be made to repeat the technique that very afternoon at the Royal Infirmary in Glasgow. The students followed their teacher to the operating theater, which was already crowded, to observe a successful operation under ether.

The French were not quite so quick to embrace etherization. The surgeons there ignored the first reports from the United

States. As the lofty Velpeau put it, he "politely declined" to experiment with ether throughout the first month that he knew about it. By January, though, the promises made for etherization were substantiated by reports from England. Also, one of Morton's inhalers arrived from Boston. Both developments encouraged trials that proved etherization not only a success in Paris but, as Velpeau graciously conceded, "a glorious conquest for humanity."

In Philadelphia, on the other hand, they were calling it a quagmire of quackery. They were also calling it a will-o'-the-wisp, and worst of all—because both the italics and quotation marks are theirs—a "*patent medicine.*" It would be July 19, 1847, almost ten months after Ether Day, before ether was used in a recorded operation in Philadelphia. There is anecdotal evidence, however, that it may have been tried earlier. A doctor named James Darrach recalled long afterward a time when a man was brought into the hospital with a crushed leg; a senior surgeon named Norris made plans to amputate and told Darrach to bring some ether. "We will try what they are doing up in Boston," he said. Darrach administered the vapor off a towel, and it soon knocked the patient out. "Dr. Norris!" Darrach cried out in alarm. "The patient is unconscious!"

Up in Boston they knew that that was precisely the idea, but Dr. Norris shared Darrach's anxiety. "Take that damned stuff away," he shouted.

In New York the great Valentine Mott could not be dissuaded from trying the new process in an operation December 8, 1846, taking care to use Letheon procured through William Morton's local agent. A month later, though, a medical journal published in New York, the *Annalist*, was as emphatic as ever in its rejection of etherization. That it was engaged in a losing cause is obvious now. That its stance was simply wrong, in view of the agony that ether, patented or unpatented, could vanquish is just as obvious today, and perhaps even too much so. Many observers would say that to the *Annalist* and regional journals all over the United States, the problem with etherization was not that it had been patented by a Bostonian, but that it had been discovered by one—and not a New Yorker, a Philadelphian, a New Orleans resident, or so forth through the hotbeds of anti-ether feeling.

However, not all the critics were easy to dismiss. The fact that so many of Boston's most respected medical men were going out of their way to aid and abet a fortune hunter was genuinely offensive to many of their colleagues around the country.

To the editors of the *Annalist,* the only good that etherization had done was to revive "in many minds, in which it had too long slumbered, a juster sense of ethical propriety. . . ." They referred to Mass General's support of the discovery as, "an attempt on the part of eminent members of the profession, to bolster up disguise and secrecy," and implied that it would "taint the purity of our science, and dishonour its votaries."

A doctor in Baltimore said that the Letheon patent was tantamount to speculation in human misery. He could try to resist it, so could others all over the country, but they were the old men, the ones who talked about the purity of medical science. Whatever their age in years, they were the old men.

William Morton was the new, openly admitting, as he put it, "the motive of profit and remuneration to myself." And in December 1846, he was hearing from patients who were telling him that there was nothing wrong in speculating in human misery, as long as he was able to reduce theirs.

In mid-December he received an envelope addressed to:

Dr. Bigelow or Morton
of the Massachusetts Hospital
Boston
pm* will please forward this to where the Hospital is if
not located in Boston.

The letter inside (shown here as written, with only a few added punctuation marks) was from a man out west:

---

*Postmaster.

Gentleman I have been reading in the publick prints some important operations by stupifying the patient by taking drugs. as I am one of these unfortunate subjects by accident, I have concluded to write you relative to my case. I have hardly means to come so far to get relief but would if you think I am a fit subject my case is this on the 8 of June last while in the Rocky Mountains 900 miles from Medical aid my mule throwed me and Broke my arm which is now well & dislocated my shoulder by throwing the socket bone out which now lays between my neck & point of shoulder. from the point of my shoulder my arm has fell down rendering my arm useless I can not rase it nor have any use of it from my Elbow up I am 35 years of age ordernary nerves please address me as above whether you could stupyfy me and set my shoulder without pain on recpt of this. Respectfully Hiram Smith. Findlay, Hancock Co, Ohio, Dec 12, 1846.

William Morton's response has not survived, but by the middle of December he already had a Letheon agent in Ohio, waiting, as it were, for Hiram Smith.

11

REPELLED BY A COMMON MOMENT

Dr. Jackson explained his position in a book called *A Manual of Etherization*, which he published in 1861, fifteen years after the discovery. He didn't merely give his side of the story, he gave his view of the world and his place in it, by choosing to quote at length a letter he had written on the subject of the controversy to Baron Alexander von Humboldt, a German toasted as the "king of science" in his day.

"The circumstances were as follows," Jackson wrote to Humboldt, as he proceeded to explain that he had first etherized himself as far back as 1842, after accidentally inhaling chlorine gas. Chlorine gas is very harsh. It caused Dr. Jackson to choke, and he immediately recalled that the fumes of sulfuric ether were sometimes prescribed for relief from breathing difficulties. He used it and derived some relief. The next day, to soothe his throat, he tried ether inhalation again, falling into a dreamy state and becoming aware as he regained full consciousness that his body was benumbed. "Reflecting on these phenomena, the idea flashed into my mind," Jackson wrote, referring to the basis of ether anesthesia, which, he claimed, he "had for so long a time been in quest of."

Jackson maintained that he told at least a dozen people about the discovery in the ensuing years. "Having confided my discovery to twelve of my friends," he continued,

> most of whom are gentlemen devoted to science, and some of them physicians and dentists, I considered it safe, so far as priority of discovery was concerned. It was my intention to revisit Europe and to bring out this discovery in the great hospitals of Paris, where I felt confident I should be treated with courtesy and fairness; but I was at the time actively engaged in the Geological Survey of the State of New Hampshire; and while my Report was in press, was called upon to explore the wilderness of Lake Superior land district, for copper mines, so that I had not a month that could be spared for a voyage to Europe. Hence my procrastination.

Hence Dr. Jackson's procrastination. All over the world, people were screaming in agony, but there was something even more important in the world to which he belonged. There was hierarchy. Paris was above Boston; Dr. Velpeau was above Dr. Jackson. Baron von Humboldt was above. Warren was below, Mass General was below, Morton—he was *way* below. That hierarchy was essential to Dr. Jackson, who had once studied in Paris, and it may have represented his chance to be hailed in Paris. But it was no reason to delay the great discovery, if indeed he made it.

"Under these circumstances," he wrote, referring to his inability to go to Paris, year after year from 1842 to 1846, "I employed a dentist, a nominal medical student of mine, Mr. W. T. G. Morton, to make a trial of my discovery, in dental surgery, which he consented to do." Jackson then related the story of Eben Frost's etherization. "It proved successful," he wrote, "the patient alleging that he felt no pain. This operation took place on the 30th of September, 1846. The case was promptly reported to me the next morning."

The two versions of the events of September 1846 have never been untangled. Either man had it in him to lie: Morton, to protect his future; Jackson, to protect his past.

Being clever, each man told a story that was plausible up to a certain point. Morton could have etherized his goldfish out in Wellesley. And so perhaps he did. Jackson could have disclosed his discovery to a dozen of his buddies. And so perhaps he did. There is only one crucial point of argument in their stories: the morning of September 30, 1846. The divergence on that morning launched a revolution in science, leaving behind both the old world and the old order. That was Charles Jackson's world. It was stormed and replaced by William Morton and his modern rapacity, so appealing in the 1840s and ever since.

There have always been those who wondered why Morton and Jackson could not share the credit for the discovery — there have been many who simply ascribed joint credit to them. To do so is convenient, but it dismisses the fact that the sides they took were more relevant than their immediate reasons for taking them. Morton and Jackson should not be made to come to the middle and shake hands like good little boys, not even for the sake of convenience, not even for the sake of a certain smaller accuracy.

Charles Jackson couldn't allow a man like Morton into his world, where science was as organized as the Church, and dominated by the European notion that only masters make discoveries, and one must earn the right to be considered a master through decorum as much as through diligence—in addition to some measure of brilliance, of course. A "vulgar" man such as William Morton could not *be* in that world at all, much less be dropped onto the top of it. If he could, then it was no world: none at all.

The chasm was just as deep from Morton's side.

William Morton couldn't bring a man such as Charles Jackson with him into his new era, his democratic and splendidly greedy new era, when science would be yet another branch of commerce, organized by only one thing: competition. Jackson and his insistent snobbery had no place in that world, and if Morton had endeavored to make room for him in it, then he would have no place in it either. He had to travel fast and he knew it; he was standing in the operating theater of the

Massachusetts General Hospital demonstrating etherization within seventeen days of the emergence of the discovery on September 30. Jackson, by his own admission, waited four years, in order to position the discovery carefully in its rightful world in Paris. Morton threw it like a paper airplane into the future. It happened to land on October 16.

12

CHARLES JACKSON'S UNIVERSE

When Charles T. Jackson was about six, according to a family story related by his niece, he "laid a small train of gunpowder from the school-room fire to the school-ma'am's chair. He asked to go to the fire, lit the train, which worked beautifully, and had the ineffable joy of seeing the school-ma'am leap from her seat crying 'Lud'a'massy! I should think I was aboard a man-of-war!'"

By the time Charles was sixteen, his sister Lydia was tired of his "chemical mania," as she called it. She fretted to their older sister, Lucy, about Charles's "raising a rebellion in the kitchen with his experiments; he can gain more knowledge by studying than by trying so many," she said. Lydia blamed herself for her brother's chemical mania, since she had given him his first book on the subject. Furthermore, she told Lucy, he endangered the house with his experiments. At any rate, she said he made it dirty.

Lydia, Lucy, and Charles T. Jackson were the children of Charles and Lucy Cotton Jackson, rich and very well-established residents of Plymouth, Massachusetts—the "Old Colony"—as people of their generation liked to call it. The Jacksons couldn't trace their roots in Plymouth to 1620 and the *Mayflower.* Not quite, but their ancestor Josiah Winslow had come over on one of the subsequent ships, the *White Angel,* which landed in 1631.

Josiah's older brother, Edward Winslow, was adventurous even for a Pilgrim, being the first in the colony to go as far west as Connecticut. He also held the governorship whenever William Bradford declined to stand. One of Edward's descendants built an impressive Georgian brick house in Plymouth in 1754, but lost it after the Revolutionary War, being unable to pay his debts since he was a Tory. It was then purchased by one of Josiah Winslow's descendants—the elder Charles Jackson's cousin. Mr. Jackson inherited the house from his cousin in 1813. The provenance of the house shows not only how the Jacksons got to live in just about the best place in Plymouth, but how insular was the society in which Charles and his sisters grew up.

Mr. Jackson was a merchant, with a handful of ships and real estate holdings in Plymouth. Both he and his wife were well educated. Lydia's daughter Ellen compiled a biography of her mother that included many of the stories she had heard about her grandmother. "She sang to Mother a great deal," wrote Ellen, "I also know that she spun sometimes on the great wheel. Mother often too remembered the delight she used to feel when as she opened the door coming home from school she perceived by the delicious scent that met her that 'Mother was frying doughnuts.'"

Charles Jackson's early childhood was relaxed and protected from pressures, judging from stories that his sister, Lydia, later told. Even when they do not pertain directly to Charles, they indicate the happy atmosphere in which he grew up. Lydia was given painting lessons as a little girl. "At this time," her daughter Ellen wrote,

> a truckload of sand had been thrown in the road near their Father's house and Mother seized the opportunity to make mud pies. While she was wholly occupied with them her Aunt came along and remarked "Pretty business for a young lady who is taking lessons in painting! How do you think your hands look?" "Oh! there's plenty of water in the akyduct," Mother answered, going on with her work.

Little Lydia had it all figured out: She would make her hands ladylike again by washing them in the town's aqueduct. "Oh, there's plenty of water in the akyduct," became a little joke repeated within the family for a time.

When Lucy was ten, Lydia six, and Charles five, their mother began to show signs of consumptive illness, while Mr. Jackson suffered from chronic indigestion. "Ill health took down his spirits," Ellen wrote of her grandfather, "and he was apt to have an air of disapproval which Mother thought kept her mother sad." When Lucy and Lydia were in their teens, Mr. Jackson composed a *Book of Moral Advice* for them, in his own handwriting. "I know mankind too well," Mr. Jackson wrote to his young daughters, and his words are haunting because his son would come to know mankind the same way. "I know their falsehood," the father wrote of people in general,

> their dissipation, their coldness to all the duties of friendship and humanity. I know their little attention paid to helpless infancy. You will meet with few friends disinterested enough to do you good offices, when you are incapable of making them any return by contributing to their interest or their pleasure or even to the gratification of their vanity.

The two Charles Jacksons, father and son, were brought up as gentlemen, but each entered fields that spun right past them, if what they expected was fairness. Perhaps terming them "gentlemen" amid the pace of the nineteenth century is only a kind way of saying that they were behind their own times.

When Charles was thirteen, his parents died within a few months of each other. Although a certain amount of the family fortune slipped away because Mr. Jackson had left no written instructions regarding his complex affairs, the children were assured of private incomes. None had inherited the slightest financial sense from their parents, and so they depended on responsible relatives to manage their investments.

For a time the role of the head of the family was assumed by a rather overbearing man named Charles Brown, whom Lucy married in 1820, when her brother was fifteen and attending a boarding school in Lancaster, Massachusetts. Young Charles Jackson often wrote to his brother-in-law regarding his allowance, which he had nicknamed "the needful." The need for cash may be typical enough, but in his letters, he was careful to mask his interest in chemistry and geology: That interest made him a rebel.

The new sciences promised no practical future in the 1820's. They were understood by very few people trained in the tradition of the classics and would certainly have seemed an indulgence to be discouraged by Brown, a broker in the shipping trade of Boston. And so, in the letters of his teenage years, Charles Jackson actually sounds as though he is denying some degenerate behavior whenever he writes about science to his brother-in-law. In a letter of December 9, 1823, he had to answer a direct inquiry about how he spent his free time, after a man named Fletcher, perhaps a teacher or a family friend, tattled back home that Charles was devoting his free time to science: "It is not however, as you suppose, spent in the study of Chemistry. I assure you I have not studied either Chemistry or Mineralogy since I have been in Lancaster except when I found an unknown mineral just to look to see what it might be."

"It is true," the confession continues,

> I have two chemical books, viz. Henry and Davy, the former I have not read a page since I have been here, the latter only 25 or 30. By this you may judge how far I am carried away by this interesting study, or as you seem to think, an infecting disease. I cannot account for Mr. Fletcher's speech unless he judges from my talking so much upon the science, for I am sure he never saw me look into a chemical book except once when he asked me the specific gravity of milk.

Those were the conditions under which Charles Jackson had to pretend, as a youth, that he wasn't interested in chemistry and geology.

It taught him to lie. In letter after letter back home, he made elaborate professions of his disinterest in the new sciences. If Brown and the others would make him feel as though he were at the mercy of a "mania," he would resist by coming to regard his alleged vice as the mistress to whom he owed his only allegiance.

Geology was in its own third epoch—epochs on the small scale—when Charles Jackson became obsessed with it as a teenager. The veritable founder of the science was Abraham Werner, a German professor of mining who gave the new discipline a rather bumpy name—"geognosy"—in the 1780s. James Hutton of Edinburgh responded with his own *Theory of the Earth* in 1795. His book asserted that current observations could trace a direct link with the past. Hutton's book might have ignited a wider following than ever for geology, except for one thing, which was that *Theory of the Earth* was among the dullest books ever put to paper. Fortunately Hutton had a friend, John Playfair, who rewrote the book in 1802 and excited a new generation. In the first decades of the nineteenth century, amateur geologists (there were no professionals) around the world were inspired by the book, especially the part about informed observation being a direct link to the past: to 100 million years or more. That was a heady feeling.

Any pipsqueak who knew what to look for in local formations could accurately suppose how they had gotten that way—and so could loom as a master over 100 million years. A very heady feeling. There were later eras in which teenagers were the first to perceive the thrill in shortwave radio, which let upstarts talk to one another between Sydney and Perth Amboy, or in personal computers, which let mere kids sort facts by the millions. In the 1820s most grown-ups just didn't understand that the same sort of limitless power lay in the new sciences of geology and chemistry.

As a youngster Jackson never told the whole truth about his studies in chemistry, and after a while, he probably didn't know the meaning of the phrase "the whole truth" when it applied to anything encroaching on his passion for chemistry and geology. The truth was on the shipping scales, though, for when it was time to move from boarding school back to Boston, Charles Jackson needed special crates in which to ship his gigantic collection of rock samples. He was ready to start college in 1825, but using his "slender stock of health" as an excuse, he bypassed Harvard College, where there were no undergraduate courses in the new sciences, only requirements covering recitations, curfews, prayer meetings, and so forth. Jackson determined on his own that such requirements would ruin his health. Instead he entered the medical school directly. There was no better way to study science in the United States at the time.

After graduating from Harvard Medical School, Jackson went to Paris in 1829 to finish his education at the highest level possible, centering himself at the University of France. Jackson could not be faulted in either Paris or Boston with neglecting his medical studies. He was considered a remarkable student in both places, with a retentive memory and a methodical mind. However, he could not resist drifting over to the École Royale des Mines (Royal Mining School) to hear lectures by the leading geologist in France, Léonce Élie de Beaumont.

Jackson became a disciple of Élie de Beaumont and, eventually, a personal friend. That was not easy to accomplish, as the French geologist terrorized most of his students. In 1832, though, Jackson responded to family pressure and reluctantly made plans to go back to Boston to begin his career.

Charles wrote to Lydia that as the packet ship *Sully* set sail from Le Havre for the United States, "I could not look on the shores of France without regretting that I was bidding a final adieu to the land of sciences and politics." Three years before, Charles Jackson had approached the most demanding scientific community in the world with boyish trepidation. Unlike most of his contemporaries from the ranks of the American ambitious and rich, however, he had done much more than merely pass through. He had passed into the intellectual being of Paris, finding a place within it that he had not known at home.

Charles T. Jackson, c. 1855.
(Van Pelt Library,
University of Pennsylvania)

Jackson would never return to the city again, but in his own mind he would never leave it behind: Its brilliance and its vanity had once been his, and so they would always be his.

But no sooner had the coastline of France receded from sight than handsome thoughts about the Old World were scattered into a flurry of those enthusiasms that were always in the air around Charles Jackson. If science was his passion, he left France in a riled state and like any other fool in love, he just couldn't help talking about it. Aboard the *Sully,* Jackson was delighted to find himself surrounded by passengers willing to converse with him practically around the clock. One day he gave a public dissection of a porpoise that had been caught in midocean, and lectured as he cut, noting especially the similarities between the porpoise heart and the human one.

"He is fond of listening," Jackson once said of a traveling companion on another trip, "and I of explaining, so we get along famously."

"And I of explaining"—that phrase described the beauty in the mind of Charles Jackson. He knew so much about his loves, his sciences, that he could go on and on about them. Sometimes, in responding to a question, he came up with a stunning idea, but in the whir of his thoughts, even he didn't always hear it. Sometimes, though, there was someone fond of listening, who did.

"Among the passengers, we had Mr. Rives, late ambassador to the Court of the French Monarch," Jackson reported of the *Sully*, picking favorites from the total manifest of about 110 people, "Mrs. Rives is a very amiable and accomplished lady, given somewhat to the muses. . . . We had Mrs. Palmer and her whole family courtesy of Miss E. Palmer and her 3 brothers. We had among others on board Mr. Morse, a distinguished American painter, who has been studying in Italy."

Samuel Finley Breese Morse was a native of Charlestown, Massachusetts (now part of Boston), a Yale graduate who had devoted most of his life to the study of painting. At forty-one he was fourteen years older than Charles Jackson. A pleasant and certainly an erudite man, he had large features on a wide face, with black hair and a dark complexion that was probably deeply tanned after a sojourn in Italy.

"It was at the table in the cabin, just after we had completed the usual repast at mid-day," Morse recalled in a letter to Jackson, of the turning point of their voyage on the *Sully*,

> You were upon one side of the table and I upon the other. We were conversing on the recent scientific discoveries in Electro-Magnetism, and the experiments of [André] Ampère with the Electro-Magnet; you were describing the length of wire in the coil of a magnet, and the question was asked by one of the passengers if the electricity was not retarded by the length of wire. You replied, no; that electricity passed instantaneously over any known length of wire.

"I mentioned," wrote Jackson in turn to Morse, picking up the same story,

> that I had seen the electric spark pass instantaneously without any appreciable loss of time, 400 times around the great lecture room of the Sorbonne. This evidently surprised the company, and I then asked if they had not read of Dr. Franklin's experiment, in which he caused electricity to go a journey of 20 miles by means of a wire stretched up the Thames, the water being made a portion of the circuit. The answer was from yourself that you had not read it. After a short discussion, as to the instantaneous nature of the passage, one of the party, either Mr. Rives or Mr. Fisher, said it would be well if we could send news in the same rapid manner.

It is no wonder that the minds of the passengers wandered from the abstraction of Jackson's accounts of electricity to their own various needs for instantaneous communication—at any given moment, a great many people in Europe would be fully engaged just wondering what was happening in some other part of Europe. In most cases there wasn't any choice except to wait for a man on a horse to arrive with news. And then for another one to gallop in and confirm the report of the first. With communication at that slow pace, even all-out wars sometimes lasted ten or fifteen years, while heads of state waited for accurate information with which to wage battle. Men such as Rives and Fisher and the others onboard the *Sully* were only typical if they were frustrated by the fact that news moved so slowly on a regional or international level.

It was either Rives or Fisher who piped up to ask Charles Jackson whether electricity could be used to send messages.

"Why can't we?" Morse interjected.

Jackson then explained that it could, replying just as though he had already thought of it—already thought of the electromagnetic telegraph. Morse, however, later doubted that Jackson had actually thought of the invention before, contending that "my suggestion of the possibility of conveying intelligence by electricity was episodical; it did not change the current of your remarks from electro-magnetism."

Of course, had Morse known Jackson longer, he might have recognized that as a moot point. The *Sully* could have done something really episodical, such as sink, and Charles Jackson would not have changed the current of his remarks until his mouth filled with seawater.

Jackson and Morse would later diverge widely as to how the idea for the telegraph developed during the course of the voyage. Jackson claimed that throughout the rest of the trip, Morse "followed me about the vessel asking me questions and taking notes in his memorandum book."

Morse allowed that he had "confided" a number of his thoughts to Jackson about the use of electricity in communication, but he discounted the importance of that by adding that he had made even more of a pest of himself to a few of the other passengers, as well as the captain.

When the *Sully* finally landed in New York in mid-November, Jackson was understandably excited to be somewhere, anywhere, after forty days in midocean. He hopped off and wrote to his relatives all about the journey, telling his brother-in-law Charles Brown that "my time onboard was far from disagreeably spent owing to the intelligent & polite company I was among and the literary conversation I was enabled to enjoy." He didn't mention the telegraph. As soon as his baggage cleared customs (a process of four days), he returned to Boston.

Morse hopped off, too, and he was met by his younger brothers, Sidney and Richard. He didn't need much time to tell them all about Italy, it appears, because on the walk home, he was already telling them in great detail about the idea for the electromagnetic telegraph. As soon as he could, he built several models and continued to develop the idea, though he was also engaged with his first full-time job, a professorship in art at the newly founded New York University.

For his part Charles Jackson neglected the telegraph and did little more with the new invention than broach the idea to a few of his friends, including Professor Benjamin Silliman at Yale University. In February 1833, he returned to Plymouth, ready to settle down, as he knew he must. Or at least, as his brother-in-law knew he must. Of far greater impetus, though, was a young woman of seventeen named Susan Bridge, with whom Charles had fallen in love. The two were married in February 1834.

Susan Bridge Jackson must have envisioned a quiet life in Boston with her husband, a doctor of such apparent promise. However, she might as well have married a sea captain, as it turned out, for she was to spend most of her married life keeping the house and raising children while Charles was away.

He first made an effort to establish a medical practice in the suburb of Roxbury (now part of Boston). No matter how long Dr. Jackson had spent in Paris, he still had to prove himself in an outer district such as Roxbury before trying to scale the profession in Boston. Because there were not yet professional standards in medicine, reputation was of paramount importance, and patients avoided new men, especially in bigger cities where the medical field was crowded anyway. Jackson complained that the patients felt no compunction to pay their doctor's bills, and so he found that he was living on his "needful," just as in his student days. Except that he had a wife to support, and before long, the first of seven children (six of whom would survive to adulthood). "My professional bills come in very poorly and I shall think myself fortunate if I collect one third of my charges there are so many poor people in my present district," he wrote to Lucy, "I have been provoked almost to giving up my present domicile, as we realize so little from my labours."

Geological surveys, meanwhile, had proved their worth as economic blueprints in several southern states, including Virginia. Having authored one of the earliest geological surveys conducted in North America (that of Nova Scotia) during his school days, Jackson found that he could have as much work as he wanted in that realm, and also as a consulting chemist. He couldn't make a living as a doctor, but opportunity was waiting in the analytical sciences: at least, that is what he told his family and himself, as he left for a long trip to survey Maine in 1835. "You would have laughed," Charles wrote to Lydia, "to have seen your brother there in his red flannel shirt with his pantaloons girded to his loins with a leathern belt, armed with tools of his trade and loaded with a heavy knapsack—Maine is a most wonderful State. You may go anywhere on her rivers and lakes in a birch bark canoe."

Charles Jackson was no different at thirty—protesting that he couldn't make a living as a doctor and had no choice but to turn to geology and life in a birchbark canoe—than he had been at sixteen. He did as he liked, and he liked geology. As he accepted commissions for other, ever-farther-flung surveys over the next fifteen years, he even took to grumbling that he regretted having devoted any time at all to medicine.

In 1834, soon after Charles's wedding, Lydia Jackson happened to go to a church in Boston. At the first sight of the minister, she did a double-take. According to her daughter, Ellen, "she didn't know a human being could have a neck so long." However, the minister, Ralph Waldo Emerson, impressed her even more with his eloquence before his sermon was over. Introduced afterward, Lydia Jackson impressed the Reverend Emerson just as brightly. He coined the euphonic name "Lidian," by which she was known ever afterward. The two were married in 1835.

And so, when the sides were drawn in Boston over the ether controversy, and the city's great institutions took their stands, William Morton had Massachusetts General Hospital as his staunchest ally—and Charles Jackson had Ralph Waldo Emerson.

As a writer and lecturer, Ralph Waldo Emerson would become a wildly popular idol in the mid–1800s: he was easily the country's most influential literary figure. His ideas, as old as the earth if he could manage it, seemed like change to that most-changing, most intently modern decade of them all, the 1840s, when he rose to national fame. In Boston, Emerson loomed at the center of a burgeoning culture of letters and ideas, out of which was founded the *Atlantic Monthly* in 1857.

Emerson had been raised by his mother in straitened circumstances due to the early death of his father, but he received a vigorous intellectual upbringing from his aunt Mary, a prodigious reader and a highly critical one. (In fact, Aunt Mary had been raised in *truly* straitened circumstances, being orphaned as a baby in the backwoods of Maine. As soon as she was able to fend for herself, she arranged to board with a long series of ministers' families, moving on as soon as she had read everything in each household. Deeply religious and utterly independent, she made intellectual demands on all her nephews, Ralph Waldo among them.)

Waldo Emerson followed his older brother, William, to Harvard. After graduation, they operated a girls' school in Boston, but both eventually turned instead to the ministry, which was Waldo's occupation when he met Lydia Jackson and married her. "For my part," Charles wrote his sister from the coast of Maine, "it seems that such a change is almost unreal since I remember all your former protestations against marriage." Emerson was not a famous man when he married Lydia, but as time went on and he began to earn a handsome living, he managed it securely. The Emersons and their five children lived in a roomy, gray clapboard house in Concord, Massachusetts, which became a haven for the other two Jacksons and their families too.

One day in 1834, out of the clear blue sky, Lucy Brown—Charles's oldest sister—received a letter from her husband, the man who had tried to suppress Charles Jackson's youthful interest in science. Confessing to gross misconduct in his business affairs, he announced that he was fleeing the country. There was no warning other than that, and no good-bye. She never saw him again; he lived in Istanbul for the rest of his life. To settle the claims against her husband, and support their three children, Lucy had to sell all her household goods. As soon as Lydia was settled in her new home in Concord, the Browns followed, and Waldo arranged for them to live in a house run by a woman named Thoreau, whom he knew only barely. It is a small point of history, literary or scientific, that Mrs. Thoreau's teenage son conceived of a crush on the widowed Lucy Brown, Charles

Jackson's sister, eventually trying to impress her by showing her a journal he kept. Lucy showed it to her brother-in-law, a light act of intercession that led to Ralph Waldo Emerson asking to meet the youngster, Henry David Thoreau.

Ralph Waldo Emerson was impressive in public, but he was unmatched as a family man. He served as a pillar of strength for Lucy's family, and as her protector. Later he would fill exactly the same roles for Charles.

During his working career Charles Jackson built the most advanced private laboratory in Boston and developed a wide reputation as an analyzing chemist, while writing dozens of articles for scientific journals on his observations in the lab or in the field. As a respected geologist and chemist, he might have been a likely candidate for a position at Harvard University, but he hated the place. "There seems to be a 'levelling downward,' in the college," he wrote, explaining why he was far better off on his own, "no man there is allowed to surpass his fellows in Science or learning without being picked upon."

Perhaps Jackson was right in remaining independent; his activities were as wide-ranging as were his interests. A number of scientists recall the joy of just hanging around Jackson's lab: to hear him talk, to watch him work, and, not least, to see who or what might come in the door. An old friend named C.A. Bartol recalled a day when a craggy farmer walked clutching a handkerchief filled with "those yellow blocks called iron pyrites, saying he had found a gold mine on his farm, and would not take no for an answer." Dr. Jackson heated the blocks until the sulfite contained within the pyrites began to burn. He held the smoldering rocks under the farmer's nose and asked him what they smelled of.

"'*Hell*' was the somewhat hasty reply," Bartol related, and with that, Dr. Jackson considered that the chemical analysis of the rocks was complete.

On Thursday, October 25, 1843, Jackson met two new clients, William T. G. Morton and Horace Wells, of Hartford. They had developed a process for manufacturing dental plates and required chemical certification of its properties in order to promote it. Three days later Dr. Jackson was prepared to give them a signed certificate.

The appointment may have been a mundane chore for Jackson, but it was a sort of honor for Wells and Morton. The commonly held opinion, even among people who didn't like Jackson, was that Boston had a genius and it was Dr. Charles Jackson. Those who viewed him more closely, his colleagues in his various fields of expertise, watched him with impatience, though. "Dr. Jackson did not always push his theories of geological phenomena to the fullness of conclusion," wrote a fellow geologist, "He had too many irons in the fire to do as he would with all of them."

Professor T. T. Bouvé, a friend at Harvard, was more pointed. "The truth is," he wrote, "Dr. Jackson was a man of great genius, and his intuitive perception of scientific truths remarkable; but from some peculiarities hard to comprehend, he often contented himself with enunciating what he recognized as fact without striving to substantiate it."

"He himself admitted his short comings in this respect," Bouvé recalled.

In September 1837, Dr. Jackson returned from a trip to Maine to find a letter awaiting him from Samuel F. B. Morse. It was something of a form letter, sent to a half dozen or so of the passengers who had sailed on the *Sully* five years before, asking in a friendly tone what each recipient recalled about Morse's conversations during that voyage regarding the electromagnetic telegraph "of which I am the inventor," he noted. The reason for the inquiry was the announcement in England that two partners named Cooke and Wheatstone had introduced an electromagnetic telegraph and were claiming it as the first in the world. One by one the passengers and even the captain replied to Morse's letter, affirming variously that they remembered him talking of little else, once he had struck on the idea for the telegraph. Then, about a month later, Morse received a startling reply from Charles Jackson. In it Dr. Jackson pointed out the slip of the pen by which Morse failed to give him equal credit for the invention. It couldn't, he allowed, have been anything but an oversight.

"My dear Sir," Morse shot back, "I lose no time in endeavoring to disabuse your mind of an error into which it has fallen in regard to the electromagnetic telegraph. . . . The discovery, so far as we alone are concerned, belongs to me." He then recounted the numerous times over the previous five years in which he had called on Dr. Jackson in person, hoping to experiment on the telegraph. "You were always otherwise busily and necessarily engaged," Morse said.

To Jackson nothing that had happened in the intervening years affected the fact that the concept for the telegraph had been his, and that he had mentioned it to Morse one day while at sea in the autumn of 1832. Morse countered that it was he who had thought of the concept, though he admitted that he had been inspired by general comments made by Jackson.

Jackson scoffed at the idea that a mere painter, a man whose most noteworthy accomplishment to date was a picture called *The Judgment of Jupiter,* could originate an idea based on an advanced understanding of electromagnetics. In that he seemed to have a point, though surprisingly, Morse did have a background in inventing. He held several patents in conjunction with his brother Sidney, including one for a fire pump that reached the market in 1817–18. But it was true that while Morse was considered one of the nation's finest historical painters, he was not on the list of its most astute men of science. A high rank in that domain was Charles Jackson's conceit.

In 1837 Morse undertook a long process for international recognition for his version of the telegraph, and for commercial success for it. For a half dozen years he literally went hungry much of the time, so small was his pay as an art professor and so demanding were the costs associated with his invention. In 1843 the U.S. government granted him the funds with which to build a line from Washington to Baltimore. After the first demonstration of the telegraph in 1844, the country witnessed a veritable melee, as copycat lines were strung up between any two likely points. Morse had to persevere through another dozen years of legal dispute before receiving his financial due, through his association with the powerful men who founded the Western Union company in 1855.

The foregoing paragraph is a cursory history of Morse's struggles to establish the telegraph. But then, it is worth recalling that James Reid, an early colleague of Morse's and the author of *The Telegraph in America and Morse Memorial* (1886), asserted repeatedly that his 872-page book was itself but a cursory account of Morse's tribulations. The achievement that resulted from those tribulations, the telegraph, is the invention for which Charles Jackson at first claimed "mutual credit" and later insisted upon exclusive credit. He sued Morse unsuccessfully for appropriating his invention and remained hotly bitter over the case ever after, as did his family. In 1873 his sister Lydia was still collecting notes about anyone who accorded her brother credit for inventing the telegraph.

"My brother Emerson," said Jackson once, referring to his brother-in-law, Ralph Waldo, "says he cannot give up his ideas and I answer him that I must hold on to facts." As he spoke Charles clenched his hand into a grasp, as though he were describing his glory and not his tragedy.

Charles Jackson could hold on to a fact without ever getting the larger idea. In a world of people and hurry and necessity and bother—in a world annoyingly real—originating the electromagnetic telegraph might have made him a genius. But it did not make him an inventor.

According to Jackson, only a great scientist had facts, and they made him supreme. He could never understand why a great idea, even in the hands of a historical painter—a nonentity—meant even more. So it was that Charles Jackson, having already watched the invention of the telegraph seep right through his fingers, considered October 28, 1843, as one of the worst days of his life, perhaps the very worst. Many years later, after pasting a copy of the certificate he issued on that date into his scrapbook, he scrawled above it, "Copy of certificate given to Welles [*sic*] & Morton. This was when I first saw them & unfortunately made their acquaintance."

That was the fact he clenched in his fist for thirty years.

13

HORACE WELLS IN PARIS

Apart and by themselves, Wells, Morton, and Jackson were ordinary men, but as a collision of ideas and emotions, they generated a spark of brilliance for which a whole world had waited hopelessly for a half century.

In the glimmering of early November 1846 each man thought he had what he wanted. Horace Wells was a hero in his home, vindicated of the shame of his failure at Mass General in 1844. William Morton had easy money within reach, hundreds of thousands of dollars by his own estimation. Dr. Charles Jackson was poised to become the peer of Europe's greatest men of science. None of the three ever got any closer to what they wanted, though, than they were in that November 1846.

On December 7 Horace Wells took up a pen to write to the *Hartford Courant* newspaper, and with that, he formally entered the public debate over the ether discovery. He was not motivated so much by acclaim for the discovery, but by the debate over the patent: that infuriating, pioneering patent. The point of Wells's letter was that he had given Jackson and Morton the idea for inhalation anesthesia while he was in Boston in 1845, demonstrating nitrous oxide at Mass General—the demonstration that had been labeled "a humbug affair," as Horace Wells recollected it in his letter to the editor.

Wells didn't mention how he had literally tripped over the

discovery, in the form of Sam Cooley, at one of Gardner Colton's lectures. Instead he carefully described how he had *reasoned* the discovery in the perfect isolation of his mind. Wells knew what he wanted people to think, and it wasn't that he was just lucky. He also knew what he wanted William Morton to think, and told him so in a letter written a couple of days later, on December 10. In that letter Wells claimed that Morton's patent was "nothing more than what I can prove priority of discovery by at least eighteen months."

"Now, I do not wish, or expect," Wells continued, "to make any money out of this invention, nor to cause you to be the loser, but I have resolved to give a history of its introduction, that I may have what credit belongs to me."

With that Wells let the matter drop. During the year he had turned his energies toward a whole new field: entertainment. He had always loved birds, and so he organized an exhibition of exotic species at Hartford's City Hall, charging a quarter to see it. It flopped. At the end of the year he was working with a partner to market one of his latest inventions, a manually pumped shower.

Although Wells flagged badly in his campaign for credit for the discovery of anesthetics, his claim received convincing support from a Hartford physician named P. W. Ellsworth, who wrote a letter published in the *Boston Medical & Surgical Journal.* Ellsworth gave William Morton due credit for introducing anesthetics successfully, but insisted that Horace Wells deserved sole credit for the actual discovery. Dr. Ellsworth was uniquely equipped to bear witness. One year earlier, well before Ether Day, he had written a long and rather dull article on new developments in surgery for the *Boston Medical & Surgical Journal*; in it, he referred explicitly to the fact that Hartford dentists were in the habit of using nitrous oxide to obviate pain. That reference, buried near the end of the piece and seemingly ignored even by those people who trudged through the whole article, constituted the only evidence in a medical journal of a priority in the use of inhalation anesthetics.

Ellsworth's 1845 article vouched for the fact that nitrous oxide was in use in Hartford before sulfuric ether was in use in Boston. In turn, his 1846 letter to the medical journal vouched for the fact that Horace Wells was the originator of that Hartford invention, nitrous oxide for use in inducing insensibility to pain.

Wells was not the only person to step forward, however, volunteering to be called the genius behind the greatest medical discovery of the age. The *Lancet* heard of so many in the first few months after England's own Ether Day in December 1846 that it took to calling them the "jump-up behinders."

The same was true in Paris, according to a resident who wrote about the latest events there in an 1847 letter: "Numberless communications are published from persons who knew all these things long ago," he wrote, referring to ether's uses in surgery, "20, 30, and 40 years since."

As far as William Morton was concerned, the primordial jump-up behinder was named Charles Jackson. Horace Wells made an insignificant squeak by comparison with the ominous rumble out of Jackson's laboratory. In mid-November, Jackson had taken advantage of his superior knowledge of scientific law—specifically, that law of gravity which made all judgments regarding priority travel down from Paris, not up from elsewhere. He did exactly what he had agreed not to do, in conversation with Morton's lawyer, R. H. Eddy: He wrote a long and persuasive letter, in learned French, to the Academy of Sciences of France. In it he described how he had discovered ether's ability to induce insensibility some years before, when he had used it on himself to allay the effects of his accidental inhalation of chlorine gas.

The letter and the diplomacy with which it was delivered to the academy through the hands of Paris's local celebrity, Élie de Beaumont, were deftly managed. "He wrote to the 'French Institute,'" wrote an American breathlessly of Jackson's campaign, using "Institute" as a translation for the Academy of Sciences. "His letter bore the Boston, Liverpool, and the

French post-marks, then it was sealed by the Institute, its receipt recorded and left sealed until ordered to be opened."

Dr. Jackson was, however, far less agile in deflecting criticism of his participation in the despised patent. Association with a medical patent was nearly impossible for a pure-hearted man of science to explain, but Charles Jackson tried. He said that he hadn't wanted a patent originally, but that once Morton and his lawyer, R. H. Eddy, had made it clear that they intended to make an application listing Morton as the sole discoverer, he felt compelled to sign his name to the application, to protect his right to credit. He said he didn't want to realize any money from etherization (which was just as well, because he never did), though he did fight for a hefty percentage of what moneys might be realized—only for the sake of defending his right to credit for the idea of anesthesia.

Even those who loved Charles Jackson and even those who despised William Morton found it hard to draw a straight line through Jackson's actions. He not only shifted his attitudes, before and after Ether Day, he shifted his explanations of those attitudes as time went on. And so Charles Jackson, who was far and away the most qualified to have made the discovery, was the hardest to understand or to defend. When Morton lied, which was often, it was because he was desperate for money, and that is understandable. As with many people of that ilk, his lies became almost a language that people who knew him could translate as he went along. Morton's crazy stories about testing ether on Nig, and so forth, could be discounted without discounting Morton.

Charles Jackson didn't lie often; he told the truth with daunting precision most of the time. And so, when his story went awry, out of a desperation for something deeper than money, as with the story of the patent, he was misunderstood, distrusted, and sorely resented. People might have trusted him more if had lied all the time.

The first fulcrum in the ether controversy was the one on which Charles Jackson's behavior teetered most severely: Ether

Day itself. By contrast Ether Day represented no pivot at all in the attitude or actions of William Morton. To him etherization was great—it was *great.* Even if one must always chide Morton by adding that those same letters also spelled "*money*" in his mind, the fact is that he was just as convinced of the efficacy of anesthesia before Ether Day as everyone else was afterward.

Charles Jackson, though, kept his distance from the discovery. In early October he was bickering with Morton's lawyer over remuneration; in late October he was entering his name as codiscoverer on the patent application. If the discovery were to be deemed a rich one or a beneficent one, he was positioned to have his stake in it. All through those supremely fascinating times, however, he stayed away from Morton and the etherization experiments at the dental office. On Ether Day he was far away from Mass General's operating theater. Likewise, he was otherwise engaged on the Saturday following, at Morton's second demonstration of ether anesthesia. And so, if the discovery were deemed a humbug, or even a form of murder, Charles Jackson was positioned to suffer no affiliation with it.

That duplicity was consistent with the system under which Jackson had studied in Europe. There a professor assigned experimental work to underlings, with the understanding that he would receive full credit for successful experiments while they would suffer for the failures. It was an utterly artificial means of boosting up "great men," but then, in practical terms, those are a commodity of far more use than mere genius, flawed as it is likely to be. The distance that Jackson maintained during October 1846 may have been wily, but it was a wile he acquired from the best of them, when he was studying overseas.

Jackson finally divulged his involvement publicly at a meeting of the prestigious Thursday Evening Club in John Warren's home. The wonder of etherization was a topic of warm interest at the club, where Morton's demonstration was described in great detail. "We then learned from Dr. Jackson," Warren wrote in his journal, "that he suggested to Dr. Morton the use of Ether for the prevention of pain in a patient who was to undergo a severe dental operation."

Dr. Jackson jumped up behind.

William Morton was not worried, not at that time. In fact in those early days, he freely admitted that the germ of the idea, the scientific and chemical basis of it, belonged to the brilliant Dr. Charles Jackson. The more credit he gave Jackson, the more reputable the discovery became. The more reputable, the more salable.

Ever since the friction at Mass General preceding the important November 8 operation—the first major surgery performed with the use of ether—Morton had resigned himself to the fact that the patent could not be used in order to charge hospitals for the painkiller. Humanitarian pressures bore down. But on that issue commercial ones did not: Morton figured that there was hardly any money to be made from the usual smattering of capital operations, anyway. His fortune lay in the field of dentistry, where relief was not exactly a humanitarian responsibility; it was just a business opportunity. Dental work was optional in many cases: If it were uniformly pain-free, customers would naturally close the gap between work that was absolutely needed and work that was merely wanted. The difference might double or even treble the patient load for an average dentist. Morton saw nothing wrong with that.

In January 1847 the debate over the Letheon patent ballooned, and sides were drawn in vociferous attacks throughout newspapers and medical journals. Doctors could not employ ether in the form of Letheon without risking fraternization with quacks: the specter laid down by the *Annalist* and many other outraged publications. Then something happened to funk the fight: The Letheon patent shattered, with a finality akin to that of a glass globe falling to the floor. Word spread that Letheon was just sulfuric ether with orange scent mixed in. A doctor in South Reading, Massachusetts, pointed it out publicly in a letter to the *Boston Medical & Surgical Journal,* a letter that simply underscored the irony of the whole affair. "While spending an evening a short time since, in a social party, composed mostly of literary friends, the inhalation of an ethereal compound formed a part of the amusements of the occasion," he said, then giving the recipe as mostly sulfuric ether, "The effects which it had on those who inhaled it were similar if not identical with those produced by the *vapor* now so much in vogue. One young lady, in particular, came so completely under the influence of it, that, had she had an aching tooth, I think I should have tried my skill at extracting." His letter was typical of the many who sought to topple the Letheon patent by identifying the secret ingredient—and to tackle Morton to the ground by proving that any dummy could have discovered etherization. Many a letter writer proved that point the same way the South Reading doctor did, by showing exactly how close he had come to doing so.

With the collapse of the patent, surgeons were relieved of that torturous choice between what was good for their profession and what was good for their patients. Letheon demanded that choice. Sulfuric ether, by itself, was good for both the profession and the patients. The attitude of J. F. Flagg and the Boston dentists he had organized to oppose etherization were quite typical. They had a convenient change of opinion and were blithely offering pain-free dentistry just as soon as the Letheon patent fell.

Even as the patent was losing its power with the general public, William Morton turned to a customer he was absolutely sure would respect a United States–government-issued patent: the United States government. On January 17, 1847, Morton dashed off a letter to the secretaries of war and of the navy, offering the nation's fighting forces in the Mexican War a reduced rate, only barely above cost, for Letheon and the inhalers with which to administer it.

The next day, in Paris, the Academy of Sciences convened a panel to see Charles Jackson's letter of November 13 opened. That formality tended to, the academy then heard it read aloud. William Morton had already been accepted as the discoverer in Paris, yet Jackson's letter was convincing. Divided in debates both official and unofficial, the academy moved to investigate the discovery of ether more thoroughly, before making a judgment as to who was responsible for it.

For some of those taking the long view of scientific development, the designation of William Morton as the hero of etherization labeled the whole discovery a fluke. Taking no place on the continuum of ideas, he represented nothing of American science or erudition. At least Charles Jackson was a product of American science: its enthusiasm and its best educational decisions. Given the choice—and in the confusion, many Americans presumed that there was a choice—Charles Jackson was the preferable hero.

Just as word filtered back to the United States that the French Academy was having trouble tracing the young history of etherization, word filtered through Paris that yet another American had staked a claim to the discovery—and that he was in the city at that very moment. Already weary of the "dreaded controversy" between Morton and Jackson, all Paris seemed to turn to Horace Wells, hopeful of finding the real hero.

Wells was in town to buy pictures, of all things. Having dropped dentistry, ornithology, and the shower business in that order and completely, as though they were the discarded toys of a toddler, he had embarked on a new career as an art dealer. The idea, which had come to him back in Hartford, was to go to Paris and buy reproductions of famous paintings, which he would ship back to the States. In New York he would have them framed and then sell them at auctions to newly prosperous Americans of the sort who couldn't tell a real Leonardo from a good copy, well-framed. It was not a bad plan, since the copies were labeled as such, except that Horace Wells knew nothing about art, importing, or auctions. He was to have his troubles with those things, but not in Paris. He would not have any troubles in Paris.

"Although I had no intentions of having anything to say about Ether or gas which has made so much noise of late," Wells wrote to his mother, "yet when it was known that I was in Paris, I was politely told that I was a *great* man!"

Home of Horace Wells's mother and stepfather, Betsey Wells Shaw and Abiather Shaw, Westmoreland, New Hampshire.
(W. Harry Archer Collection, University of Pittsburgh School of Dentistry)

For my name had gone before me. You are aware that Drs. Jackson and Morton have been claiming a discovery which properly belongs to me. Dr. Jackson it seems had sent out letters to Europe claiming the discovery, he therefore was first known there—but soon after an article* which was published in the Boston Medical Journal which gave me

*Dr. Ellsworth's.

> the credit of it was republished there—by the way you cannot conceive what excitement there is in Europe on this subject—indeed it is the topic of the day and is admitted by all to be the most important discovery of the age, and they are desirous of knowing who is the discoverer and it is rumored that whoever substantiates his claim to priority is to receive a pension from the French Government. Well, after it was known that I was in Paris I was invited to address the various Scientific Academies upon the subject—which I did—and not only this, several gentlemen wearing the Royal Ribbon interested themselves in my behalf and are now electioneering for me with a right good will. I was invited to parties, soirees, balls, and dinners constantly for the last 10 days of my residence in Paris; indeed balls and parties embracing the aristocracy of Paris were given in honor of me—in short I was quite a Lion, having horses and equipage second only to that of the king.

Wells told his mother that he had been invited to become the official dentist of the king, Louis Philippe. "Dr. Jackson has influential friends in France also," Wells warned her—and himself—"and he will do his best to establish his claim to this discovery although he really has no right in justice to this honor yet it is not impossible but he may outgeneral me."

Dr. Jackson, for his part, had made a miscalculation in his letter to the academy, an error that lit suspicion in the hearts of many otherwise sympathetic observers, especially American ones. He failed to mention Dr. Morton's name in any way whatsoever. That struck many people as ungenerous, undemocratic, and plain untrue. The fact was that Jackson couldn't bear to have his name connected in any way with that of William Morton: The mere sight of the two names on the same page compelled him to write many letters to many editors throughout the years, refuting any connection between himself and William Morton.

Jackson's reputation for arrogance also stemmed from his refusal to allow the question of priority in the discovery to be judged by any American court or tribunal. The only panel to which he submitted his evidence was the one in Paris. In the United States, many months went by before he could be prevailed upon, by no less than John Collins Warren, to at least explain his side of the story in a letter for publication.

To some Jackson's refusal to submit to arbitration in America betrayed his arrogance. And to some it confirmed that he had no real claim.

By March 1847 William Morton was in financial straits. Having practically abandoned his dental practice in order to manage his interests related to the discovery, he was seeing his cherished prediction of a personal profit of $360,000 drop to nothing—and then dip below that, in terms of his outlay for expenses.

Morton was not only in the Letheon business, though; he was also in the inhaler business, having ordered fifteen hundred of them, built to specification, from a manufacturer. Even as he did so, unfortunately, word was getting around medical circles that the Morton inhaler was part of the problem in consistent ether administration. A sponge soaked in the liquid became the preferred method for a time, though it was eventually replaced by a cone-shaped inhaler.

The jolting end to Morton's dreams of business profit was not *even* the news from Washington that the Departments of War and the Navy had declined his offer of a cut rate on Letheon. It was the news from Mexico that U.S. military surgeons at the front were using sulfuric ether in blatant disregard of the patent.

Even the U.S. government was violating the patent—its own patent.

The United States was fretting its way through its first long-distance conflict, the Mexican-American War of 1846–47. Washington couldn't even command its own generals in the Mexican interior; it was in no position to scold the military surgeons mopping up in their wake.

In talking about anesthetics—especially before they were invented—some hardhearted people noted that soldiers injured in battle did not require painkillers, but rode along stoically with even the most gruesome of injuries. Soldiers and sailors were buffered from agony not by drugs, but only by the intensity of battlefield emotions and the near hysteria that resulted.

The ability of the mind to distract itself from brutal reality is part of the mechanism of all consciousness, with its levers of feeling and of control. That a type of self-hypnotism—the lapse of feeling in favor of control—occurs in battle is no doubt true. But then, anything can happen in battle, in terms of those twisting pulses of feeling and control. Only in retrospect, however, would it seem that soldiers didn't require help.

William Beaumont, an army surgeon during the Revolutionary War, recorded a scene that could have occurred in any battle from the dawn of history to the present. He was at the Battle of Little York (in New York State): "A most distressing scene ensues in the Hospital—nothing but the groans of the wounded and agonies of the Dying are to be heard. The Surgeons wading in blood, cutting off arms, legs and trepanning heads to recuse their fellow creatures from untimely deaths."

Dr. Beaumont recorded the sounds of the camp hospital, as well as the sights:

> To hear the poor creatures crying, "Oh Dear! Oh, Dear! Oh, Dear! Oh, my God, my God! Do Doctor, Doctor! Do cut off my leg, my arm, my head, to relieve me from misery! I can't live, I can't live!" would have rent the heart of steel, and shocked the insensibility of the most hardened assassin and the cruelest savage. It awoke my liveliest sympathy and I cut and slashed for 48 hours without food or sleep. My God! Who can think of the shocking scene when his fellow-creatures lie mashed and mangled in every part, with a leg, an arm, a head or a body ground in pieces, without having his very heart pained with the acutest sensibility and his blood chill in his veins. Then, who can behold it without agonizing sympathy!

Not the surgeons in Mexico with the American generals: Ether was used in the Mexican-American War without regard to the patent.

The patent crumbled, and there was no one to sue over it, because there was *everyone* to sue. Not only were surgeons and dentists throughout the country beginning to use ether in place of the patented Letheon, but the very government charged with upholding the patent was the greatest infringer. Eventually William Morton would put a price tag on his involuntary contribution to medicine. For the time being, however, in early 1847, he at least accepted the fact that Patent 4848 offered no protection for the discovery of etherization. Perhaps it should never have been issued in the first place.

He had received flimsy legal advice. A known substance, such as sulfuric ether, could not meet the legal definition of an invention or design. A more discerning lawyer might have advised Morton that patent protection, even if it were granted, could not hold. A more dispassionate lawyer might have told him that retaining commercial rights over a common substance—cornering the market on it—was out of the question. But then, William Morton was no more rooted in the legal world than he was in the medical one.

Officially the patent still stood. It remained in the government record. It implied certain rights. In practice, though, no one in the private sector respected those rights by giving Morton a royalty payment. The most obvious course, normally, would have been to petition the court for protection. If he did that, however, the judge just might reject the validity of the patent. That would render it worthless. As it was, the patent had value, if not strength. That may have been a fine line, but it showed Morton a path.

Since the patent was unenforceable anyway, Morton withdrew from skirmishes with infringers from below, as well as negotiations with hospitals from above, leaving them to fall all over one another in the mad rush to etherization.

But there still had to be a way of making money on a good

idea. That was the impetus, the battle cry, of William Morton and the age he arrived with.

The collapse of the patent represented another landmark in the ether controversy, the one that helps track William Morton's shifting behavior. Before that he was less concerned with claiming credit; indeed, he was inclined to hand it over, if it gave etherization a better reputation. After the patent collapsed, however, Morton retreated from his original cash-and-carry attitude about medical discovery. The accepted formula in the old world of science was that reward was not to be grabbed at, it was to be bestowed.

Morton went about the next stage his own way, the way of a con man: He descended on the U.S. government and tried to coerce it into bestowing an award of one hundred thousand dollars on him, in exchange for the patent. From that point on, he claimed full, sole, and glorious individual credit for the discovery of etherization.

The U.S. government was rather startled by the Morton siege. It had never bestowed much money on anyone up to that point. In February 1847, though, a select committee of the House of Representatives debated the controversy and voted to recommend Morton's claim for an award in lieu of lost patent revenues. The bill was sent to the full House.

For William Morton prosecuting the dreaded controversy became a full-time job. Dr. Jackson, however, continued to pursue his career as a geologist, accepting commissions for geological surveys. Within the scope of recent human history, the mid–nineteenth century in the United States was an especially rewarding time to be charged with exploring the land and its contents. Jackson received his most promising assignment in that regard in 1847, when he was given charge of a major survey of the Lake Superior region and spent most of that year in the steep hills of the lake's western shore.

Even while Horace Wells was away in Paris, with his liveried carriages and *grands bals,* he gained a benefactor as potent in its way as Morton's loyal Mass General. His benefactor was Connecticut. Leaders in the state who didn't know much more about the story than what they read in the papers could see that Horace Wells was ill-equipped to take on the likes of Morton and Jackson. He lacked any connection to an institution and had never been associated with any successful idea or venture. He was an intelligent man of good character, but he wasn't aggressive, he wasn't a fighter. And as he was so obviously in for a fight, the elected officials of Connecticut took his side: for justice and for other, more self-serving motives. At the beginning of March a U.S. representative from Connecticut spoke out against the select committee's preliminary vote in Morton's favor, arguing successfully that it should wait to hear evidence to be gathered in Horace Wells's behalf.

At the time, Wells was blithely oblivious of the representative's arguments. In March 1847 he was on his way back to the United States, having been in Europe for about two months. "He was my fellow passenger on board the steamship *Hibernia*," recalled a Massachusetts man named William T. Davis almost sixty years later, "and shared my stateroom."

"He was a landsman," Davis said delicately of Wells, by way of explanation for what was to follow, "unfamiliar with the sea and easily frightened by the noises of the ship. He was especially frightened on a dark night in a northwest gale surrounded by broken ice." The *Hibernia* was continually shaken by collisions with blocks of ice. That night Davis and Wells were in a lounge when they heard the order "hard a' port" given to the wheelman. A moment later they heard the opposite order, "hard a' starboard," called.

"I saw on the port side perhaps a quarter of a mile distant the glisten of an iceberg," wrote Davis, "and those on the starboard side saw the glisten of another about the same distance away, and as we went wallowing along in the trough of the sea we sailed between them. We turned in soon after, but there was not much sleep for the poor Doctor [Wells] after the fright he had received."

"About midnight," Davis continued,

> [W]e were awakened by the crash on our decks of a gigantic wave, which enveloped the ship, filling the dining saloon sill deep, and pouring down into the cabin, endangering the lives of several passengers whose stateroom doors were broken open, and who were washed out of their berths.
>
> The Doctor was out and off in an instant, returning in about ten minutes telling me to get up as the ship was sinking.

Davis remained in his berth, despite Wells's entreaties: "either taking no stock in his outcry," as he said, "or thinking that a speedy death in my stateroom would be better than a lingering one among floating cakes of ice." Nonetheless the *Hibernia* landed safely in Boston.

Horace Wells never went to sea again. He returned to Hartford and immediately set about gathering affidavits and testimonials—the endless ammunition of the ether controversy. The House of Representatives had formally requested evidence of Wells's claim, and the academy in Paris was also awaiting written support.

In March, Wells met with Morton's representative in Boston, Edward J. Warren. A young businessman originally from Maine, Edward Warren was only distantly related to the surgeon John Collins Warren—however, any confusion arising from the shared surname was only to William Morton's benefit. Edward J. (as he will be called herein, leaving only one "Warren" in the story) was extremely active during 1847, initially as Morton's business partner and then, as their licensing prospects faded and were replaced by pursuit of an award from Congress, as his de facto campaign manager. Edward J. wrote a stream of letters to the press on Morton's behalf, cowrote and distributed pamphlets with his champion, and attended an increasing number of meetings with political figures, in hopes of securing early support for the Morton petition in Congress. At some juncture, probably at the first meeting between Edward J. and Wells, the Morton side offered to purchase all of Wells's rights to claim the discovery. Such an arrangement might well have been acceptable the previous November, but by the time Horace Wells returned from Paris, he believed in himself. He had been "politely" told there that he was a great man.

And so, in April, Edward J. began a campaign in the Boston papers (which were read and reprinted internationally on the subject of the etherization controversy) undercutting Wells's claim. Among other things, he brought out the note that Wells had written in response to Morton's first letter about the success of etherization, and pointed out that Wells exhibited in it no particular knowledge of anesthetics. Wells answered that attack, and also one written by Morton himself, for the *Boston Post.* "My reply will knock him into a cocked hat," Wells wrote to his brother-in-law Joseph Wales (Elizabeth's brother).

Wells had a valid point in his contention that he had discovered inhalation anesthesia by his experiments with nitrous oxide. It was weakened in two respects, though: first, because sulfuric ether was the anesthetic in the news at the time. Though Wells's use of nitrous oxide in dentistry laid out a direction for the discovery of etherization, that was too fine a point to carry very far on its own. Second, Wells's claim was weakened by the fact that he had failed to follow through with his initial discovery. During the summer of 1847, he tried to make amends on both of those fronts by administering nitrous oxide in medical operations in both New York and Hartford. Several Connecticut surgeons made use of the gas in operations, in a spirit of loyalty to the local hero, but it did not catch on.

Nitrous oxide is preferable to sulfuric ether because it has a milder effect on the system; however, great expertise must be brought to bear in order to induce and prolong a suitable level of insensibility for surgery.* Neither Wells nor any of those game Connecticut surgeons of 1847 could pretend that nitrous

---

*In modern times nitrous oxide has been used extensively during surgery, as a precursor to more potent anesthetics.

oxide, all by itself, was an improvement on ether in major operations. Utterly cast aside, nitrous oxide was not even used in dental cases after the advent of etherization. (Its return to dental practice came in the 1860s, when Gardner Colton reintroduced it and subsequently earned a tidy fortune, administering it for dentists.)

The lack of medical interest in nitrous oxide during the late 1840s made Wells's case rather steeper. He could not claim to be the *practical* hero of the discovery, since his entry, nitrous oxide, had been a dead end. However, as events unraveled, the fact that he had not followed through on his original discovery turned completely around from a negative to a positive. In the midst of the nasty dogfight that followed in the wake of Ether Day, Horace Wells could proclaim himself to be the only one of the three claimants who had never sought personal gain by the discovery. Compared to the rapacity of Morton and Jackson, Wells's very complacency cast him with dignity.

One might imagine, in a fictionalized version of events, Horace Wells accepting his part—as only a part—of the discovery of anesthetics. None of the three should ever have claimed any more than that—a part. Oddly enough, each of the three had a secondary role. What had the primary role? The air in between, which is the very meaning of "inspiration," taken from the Latin root.

In May 1847, however, the Connecticut General Assembly passed a resolution recognizing Horace Wells as the sole discoverer. It was out of his hands. Ether Day had yet another great man.

14

CHLORY

A young doctor, Milan Carey, wrote to a former classmate about the new anesthetic in April 1847. "What do you think of the inhalation of 'Sulphuric Ether,' in cases requiring operations?" he wrote. "I hope you are not among those who have no faith in it, for sure I am that no one doubts its efficacy who has seen its operation a few times."

There were those who rejected etherization all the more vociferously because they had no doubt that it worked. "Pain is the wise provision of nature," shouted a physician at a public meeting, "and patients ought to suffer pain while their surgeon is operating; they are all the better for it, and recover better!"

To a clergyman writing to Dr. James Simpson of Edinburgh, etherization was nothing but "a decoy of Satan."

"In the end," he predicted, "it will harden society and rob God of the deep earnest cries which arise in time of trouble, for help."

He may have been right at that. In one stroke anesthetics stole a victory for science, that brash religion of the mind. It left behind all those religions that fell in on just such mysteries as pain, by speaking directly to the heart or soul. Perhaps that did help to harden society, being part of a process marked by a hundred other fantastic breakthroughs in science, in the hundred centuries since civilization began. It is hard to believe that anesthetics, though, had anything to do with Satan. Many of the people who made a dilemma of that point were interested specifically in the use of anesthetics in childbirth, wherein, some considered, a mother's pain was mandated in the Bible.

When word of the demonstration at the Massachusetts General Hospital reached Edinburgh, James Simpson became its vociferous champion. Simpson, a burly man with wavy hair curling past his collar, was a specialist in obstetrics. In January 1847, when he was appointed one of Queen Victoria's surgeons, he wrote to his brother, "Flattery from the Queen is perhaps not common flattery, but I am far less interested in it than in having delivered a woman [in childbirth] this week without any pain while inhaling sulphuric ether."

"I can think of nothing else," he told his brother, regarding the potential of etherization.

By the summer of 1847 Simpson was still obsessed—not with ether, specifically, but with the idea that inhalation was indeed the best way to deliver an anesthetic into the bloodstream and so to the nervous system. Ether was flawed, mostly because it smelled, according to Simpson, but also because it was highly flammable in a day when flame was the only source of artificial light. Besides, the amount required to render a person unconscious tended to make him or her sick to the stomach. Simpson felt sure that some other chemical could fulfill even more effectively the potential that ether had revealed.

James Simpson must have been mindful of one last flaw in the ether inhalation technique, which was that it had been developed in the United States. Many Britons came right out with regret about that fact, wondering indignantly why such a superlative discovery wasn't British. James Simpson was a man of pure scientific motives, but he couldn't fail to be aware that many of his countrymen were yearning for a new substance, only to right the relationship between the mother country and the United States.

Simpson gathered a team of three other scientists to conduct research on the compounds available. Every Thursday evening

Sir James Y. Simpson (1811–70).
(Wood Library-Museum, Park Ridge, Illinois)

they would gather at the Simpson home, sitting around the dining table to inhale candidate chemicals. "I selected for experiment and have inhaled several chemical liquids of a more fragrant or agreeable odor," Simpson wrote in a medical journal during the course of his research, "such as the chloride of hydrocarbon, acetone, nitrate of oxide of ethyle, benzin, the vapour of iodoform, etc."

One old friend, a professor named Miller, made a habit of dropping by at breakfast time every Friday, so he said, to see if anyone was dead.

Without too much else to depend on, such as an understanding of the workings of anesthetics (which only barely exists even today) Simpson's prime criterion in selecting chemicals was their volatility, or tendency to evaporate, as does ether, at room temperature. Volatile liquids were the most likely to insinuate themselves into the bloodstream through inhalation, Simpson thought. On November 3, 1847, he turned to chloroform, a compound of hydrogen, carbon, and chlorine that had first been produced sixteen years before in three separate incidents only months apart. In unrelated experiments, scientists in the United States, France, and Germany each produced the new liquid, which is clear and relatively heavy, with a faintly sweet smell. None of them knew what to do with it, exactly, but since it ended up on a list of "volatile" liquids, Simpson duly placed it on his own list of possible replacements for ether, and he ordered a vial of it from a local pharmacy. On Thursday, November 4, however, there was a mad scramble after he suggested his team try it: Simpson suddenly realized he'd lost the vial. Someone finally found it under a pile of wastepaper in his home office.

On sniffing chloroform with the others, Simpson suddenly found himself transported from his dining room to a cotton mill.

As he awoke, he realized that he was lying under the dining table. He slowly recognized that what he had taken for the sounds of the cotton mill were the deep snoring of one colleague and the attempts of another to kick the table apart. "This is far better and stronger than ether," Dr. Simpson concluded, even before he got up from the floor. He was not far from the truth: The compound was not flammable, it was highly potent and predictable, and it was inexpensive. With the famed Scottish doctor promoting it, chloroform soon became the most popular anesthetic in Britain. It then spread quickly throughout the world, over the route prepared by etherization. Chloroform was considered an improvement because it worked consistently in small doses—and it didn't have that sickly sweet smell that made ether so obnoxious for those who worked with it daily.

Even as Dr. Simpson was pioneering the new compound, his greatest interest remained in the use of anesthetics to relieve the pain of childbirth, a campaign that continued to draw bitter resistance on religious grounds. When fundamentalists protested that childbirth was ordained by certain biblical references as a natural and necessarily painful process, Simpson answered with his own close reading of the Old Testament, pointing to passages showing God's mercy regarding the pain of life's transitions, including birth. Simpson's attempt at Bible study failed to move the debate, however. It was Queen Victoria, looking a little bit like God to many people—at least from a distance—who settled the whole question by requesting chloroform for the delivery of her son Leopold in 1853.

James Simpson was acclaimed for his suggestion of chloroform as an anesthetic. On being knighted, he chose for his new crest the motto *Victo Dolore*—pain conquered. Sir James was no doubt dedicated, inasmuch as that word indicates that he was focused, energetic, and even brilliant. Sir James was also reckless. A friend in the sciences, Lord Playfair, recalled that Simpson soon became disenchanted with chloroform, or "chlory," as he liked to call it, and started an inquiry into potential replacements. After some research Playfair suggested bibromide of ethylene. Sir James no sooner heard the words than he wanted to try it on himself, but his wife interceded, asking Playfair to try it first on rabbits. The following day Sir James returned to Playfair's lab and declared himself ready to inhale the bibromide. Mrs. Simpson asked to see the rabbits first. An assistant went into another room to check on them. "When the attendant came in, we saw him holding by the ears two rabbits," recalled Playfair, "perfectly dead."

James Simpson never did find a substitute for chlory, which was to remain the predominant form of anesthetic for surgery until after World War I (when a team of British chemists, specifically charged with finding a substitute, produced cyclopropane). Sulfuric ether was a perennial second choice; it remained in common use in American hospitals until the middle of the twentieth century. What is remarkable is that no viable third choice emerged for so long after the first flurry of development in anesthetics from October 1846 to November 1847.

Nonetheless James Simpson continued his active search for a third choice, because chloroform had one terrible drawback: Sometimes it killed people. Death from ether, on the hand, was practically unknown, except when a very basic rule was broken and common air was not mixed into the fumes. Overdoses during etherization were rare. Chloroform was far more dangerous. Surgeons preferred it because it acted quickly, with few side effects (such as the nausea associated with ether). "As every army surgeon can testify," so testified an army surgeon who favored chloroform, "the sufferer has been narcotized, subjected to the amputation of a limb, the vessels tied, and the stump dressed within the time usually requisite for the full effect of ether."

Dr. Simpson even suggested in an early pamphlet that the liquid, with its added advantage of being easily transported, could be issued to soldiers going into battle, so that they could use it on themselves if they were in pain and waiting for help. Ideas for the use of chloroform proliferated.

In September 1848, a political journal printed an editorial supporting the administration of chloroform to prisoners about to be hanged, so that they would not know the horrific pain of death on the gallows. The temptation to use it as a form of euthanasia was great, and it was in fact used for that purpose on sick animals.

In England the decision was made to destroy a 120-year-old elephant that could no longer walk; at a time when the humane treatment of animals was becoming a widespread priority, especially in Britain and United States, the owner specified the use of chloroform. Sadly enough, the chemical, even in an enormous dose, failed to kill the elephant. The animal was shot in the neck, the usual method of destruction, and a fountain of its blood shot into the air. Even then the animal's great spirit refused to yield. The owner and the veterinarian grew frantic, having so hoped to spare the animal just such a disquieted death. At last the vet slit the carotid artery in the neck with a long gash, and finally, according to the report that ran in newspapers around the world, the huge animal ceased to live.

"One of the happiest uses to which chloroform has been applied," wrote a military physician named Charles C. Bourbaugh, "is in the suppression of malingering." Inductees who claimed to be deaf, mute, lame, or in any other easily pretended state were given a dose of chloroform before they were sent back home. "The mute have been betrayed into vociferous speech and the deaf been made to hear," noted Dr. Bourbaugh with satisfaction, in recounting the impostors whom chloroform had put back into the army's ranks. "Rheumatism disappears as if by magic," he gurgled.

The vogue for chloroform led to a further use—if abuse is nonetheless use. After legitimate researchers and patients proclaimed that chloroform was the source of immense pleasure, however temporal, an interest in the newly famous chemical awoke in those people ever needful of immense pleasure, or even those pathetic souls simply curious regarding any pleasure at all. Chloroform was not the first chemical to attract a hedonistic following. The other anesthetics had, after all, been brought forth to kill boredom, not pain, before Ether Day, 1846. However, it took a dedicated sort of a person to become in any sense addicted to nitrous oxide in the mid–1800s, since making it was tricky and storing it practically impossible.

As to becoming addicted to the inhalation of sulfuric ether, that took unusual dedication for the same reasons. James Graham, an ether addict in London, solved the problems by becoming very good friends with his local pharmacist, whom he thanked in a book inscription in 1790 for refusing to give ether to Graham himself (writing in the third person) "for the purpose of immoderately sniffing it up his nose and thereby affecting his brain." James Graham was a handsome figure of a man, so it was said, who parlayed his personal magnetism into a splendid run as London's leading quack. Graham didn't claim to be able to cure just anything, exactly. His specialty was infertility, and for that he prescribed use of the "celestial bed": twelve feet by nine, supported by forty glass pillars of different colors, surrounded by magnets, and filled with herbs. If all of that were not enough, there was the cost, which was even steeper than the pillars, and probably more inspiring.

How James Graham managed to spend the fortune he made from the celestial bed isn't known, but somewhere in the adventure of doing so he discovered sulfuric ether. As a man grown very old by his forties, he died an addict. The poet Robert Southey, who later partook in Laughing Gas with Humphry Davy, Samuel Coleridge, and their crowd, knew Graham in London in the early 1790s, and reported that "he would madden himself with ether, run out into the street, and strip himself to clothe the first beggar he met."

To be an ether addict was to be best friends with a chemist or pharmacist, though, because the volatile liquid did not store well. Indulgers in Ireland solved that problem: They used it up immediately by drinking it. Ether drinking became so popular that it was said without exaggeration that a traveler could smell certain towns in the north a half mile away. Inebriation didn't last long, and so it barely interfered with chores and other activities. The physiologist Benjamin Ward Richardson put it more puckishly, saying of the ether drinker, "He may become a savage, but never becomes a sot." In the short run, there were only two real dangers. The first, albeit, was death, which could result from overdrinking. The second was setting one's mouth on fire. That happened when somebody lit a pipe while ether was still present about the lips.

One attribute of ether drinking that impressed those physicians who did study it, including the British authorities Norman Kerr and Benjamin Ward Richardson, is that it did not create in users what was then known as a "morbid craving," or what we know so commonly today as an addiction. Until the late nineteenth century, addiction to any substance except for tobacco and alcohol (and the herbal liqueur absinthe) was rarely known in Europe and United States, though the use of other substances around the world was familiar by name to those in the West. Among the well-known intoxicants were cocaine in South America, hashish in the Middle East, and opium in Asia. There were others, however, as almost every corner of the earth brought forth some natural element, some toggle of poison and pleasure: amanitine in eastern Siberia and red thorn apple in the Andes; arsenic in the mining region of southern Austria.

In Western Europe and the United States, chloroform represented the first dangerous temptation to come from the world of chemistry. In Western civilization, one of the strongest currents of the 1800s was the temperance movement. Its success in changing attitudes about drunkenness was considered by many who were alive then as the signal achievement of the whole century. Any comparison of the mores of 1801 and 1901 would reflect a change from a time when the subject of who was drunk after dinner was no more interesting than, say, who ate dessert, to a time when drunkenness was, at least, regarded as shameful. Yet the need to be drunk—whatever that means, as a chemical or emotional mechanism—did not decline with the reputation of alcohol.

"I had always been temperate—almost a total abstainer, in fact," wrote a Detroit patient to his doctor in the 1880s referring to alcohol use. He wrote on the understanding that his name would never be attached to the story he was about to write. "With me," he continued, "the chloroform infatuation was a case of love at first sight." He described himself at the time as a student, "fond of study and taking a keen interest in everything about me." With some curiosity about the feeling of sleep under chloroform, he brought a small bottle home, inhaled it just before he went to bed one night, "and felt," as he described it,

> the delightful sensation of being wafted through an enchanted land into Nirvana. Those who know nothing of intoxication except in the vulgar form produced by whisky, have yet to learn what power there can be in a poison to create in a moment an Elysium of delight. It is a heaven of chaste pleasures. What I most remember is the vivid pictures that would seem to pass before my eyes—creations of marvelous beauty—every image distinct in outline, perfect in symmetry and brilliant in coloring. The enjoyment is purely passive; you have only to watch vision after vision, but why each vision seems more wonderful

> and charming than the last you cannot tell, and you do not stop to question.

The Detroiter didn't. "Before I knew it I was a slave," he wrote. The chloroform betrayed him, though: He no longer had the visions that had once seemed so wonderful. Every few weeks, he would, as he described it, "indulge in a regular spree, lasting from one to three days, during which I would keep myself as nearly as possible dead drunk." By "dead drunk" he was referring solely to the influence of chloroform. Finally he was found in what appeared to be a lifeless state and the doctor to whom he later wrote attended him. After two years of attempts and failures, he finally stopped using chloroform.

Before Ether Day there was alcohol, in its many forms. Afterward, there was a new chemistry, and artificial ways to reach into the mind. Ether and chloroform started surgical science toward an ascendancy that has yet to pause. Chloroform also gave a dawning to a modern world of addiction.

"I had no special cares to drown," the Detroiter wrote of his need for chloroform intoxication,

> but it became my tyrannical pleasure to draw over my senses the veil of oblivion. I loved the valley of the shadow of death. I knew there was danger that some night I should pass over the line, into a sleep from which there would be no waking, but death held no terrors for me. Nay, to bring all of my faculties and powers and ambitions into the sweet oblivion of transient death was the one pleasure for which I cared to live.

On the way to New York in late fall 1847, Horace Wells stopped in New Haven and visited the offices of the newspaper there, a formality expected of the famous in those days. The editors received him with interest but reported that he seemed ill at ease. Elizabeth Wells also noted her husband's despondency at that time. She ascribed it to a sense of fear and despair "that the fame due him would not be accorded after his death." There had been progress on his behalf, but in some odd way he wasn't a part of it. The report that the Connecticut legislature had officially credited Horace Wells as the discoverer of inhalation anesthetics had made something of a personage of him. The state's resolution was important news across the country and in the capitals of Europe. For the first time a government body had passed judgment on the etherization controversy; that alone made it news. A gauntlet had been thrown down, and by an old, influential state. In more recent times, state legislatures have shown a willingness, even an enthusiasm, for votes on peripheral issues—such as the designation of an official state cookie. In the mid–nineteenth century, however, the intervention of a state government in a nonpolitical issue was neither common nor easily dismissed: especially not by the officials within the state, who took up Wells's cause as a mission.

In truth the emergence of Horace Wells and his eloquent defenders was a godsend for all those in the U.S. Congress who just didn't want to give one hundred thousand dollars to anybody for anything, least of all to Morton for Letheon. A tightfisted Congressman could stall the petition regarding that award simply by granting due respect to the members from Connecticut.

By the autumn of 1847, Horace Wells had involved himself once again in dentistry, going into partnership with a respectable Hartford dentist named J. B. Terry. However, it appears that he did not take an active part in the practice; he only lent his name to it while he continued to pursue his interests in selling reproduction artwork. He made several trips to New York, where he had placed an order for frames to fit his pictures. Wells then sold the works at auctions. A receipt still exists today, covering the sale of about ten paintings to one customer at an auction in Hartford; the average price was eight dollars per picture.

Medical science was jetting forward, but without Horace

Wells. He was selling pictures. To uphold his claim as the founder of the new science, the great man at the core of anesthetic development, he tried to catch up and fashion himself into an expert on chloroform. Wells experimented on himself, though, and it was chloroform that became an expert on him, his every strength and weakness.

15

ALL ALONE AT THE TOMBS

A dozen rather too gorgeous women converged on New York's main police station on Saturday, January 22, 1848, some followed by their black maids and all of them followed by a small crowd that included anyone in the station house with nothing better to do. One of those in the trailing throng was a reporter for the *New York Herald.* Other papers later described the protesters as "women of fallen character," but not the *Herald.* In its report they were simply "good-looking, well-dressed females . . . who promenade Broadway in the evening."

They came to the station well armed with evidence, or at least their maids did, carrying armloads of clothing and hats of the rarest quality, burned through with holes. Jane Montgomery was the very first to testify, telling a judge named Osborne that she had been walking on Broadway—"patrolling," as one paper had it—the previous Tuesday night, when a man hurled vitriol at her. Vitriol, another name for sulfuric acid, is strong enough to eat easily through cloth, and it can blister and even rupture skin. Next, Julia Meadows, Louisa Johnson, Maria Taylor, and Mary Pierce stepped forward in turn to say that their clothes had also been vandalized in vitriol attacks Tuesday night. The women had remained on guard over the following nights. And they were enraged. They accepted the fact that vitriol was a common tool of revenge in specific cases, but were outraged

Horace Wells, c. 1845—his only son, Charles, kept this painting in his home all his life.
(W. Harry Archer Collection, University of Pittsburgh School of Dentistry)

that it was being used wholesale on innocent strangers: all of them women, all streetwalkers.

On Friday just after nine o'clock, the night before the hearing, Jane White had been walking near the Astor House hotel on Broadway with her friend Louisa Mariad. Suddenly, she heard the sound of liquid splashing in a container. As she turned toward the noise, a man in a loose-fitting coat brushed past and she felt a sizzle of pain on her neck. "There!" she cried out. "There's the man that burns the girls with vitriol!" She lunged at him and detained him as best she could, with help from her friend Louisa, until a policeman named Beard arrived and took charge. The attacker said his name was Jonathan Hall, gave his age as thirty-five, and told the officer that he was from Charleston, South Carolina. On being pressed, however, he conceded that "Hall" was an alias, that his real name was "Smith." Officer Beard brought Smith to the New York jail known as the Tombs, and the next day Judge Osborne presided over an arraignment in the police station there.

Osborne heard testimony from Jane Montgomery and the other women one at a time, describing vitriol attacks received on

Tuesday and Friday nights. One victim, however, was unable to testify in person. According to the others she was so badly burned around the face and neck that she was confined to the hospital and had little chance of recovering without severe scars—"if she recovers," added the *Express* ominously, in its report.

On the charge of vandalizing clothing valued in the hundreds of dollars, Judge Osborne set bail at two thousand dollars. However, for maiming the young woman who was said to be still in the hospital, the prisoner was to be held without bail until her recovery was assured. Osborne seemed to concur with the description of the attacker made by the *Tribune*'s reporter: "This monster," it called him. "This monster in human shape."

It is hard to know exactly what Horace Wells was doing in New York City in the third week of January 1848. By the time he found himself sitting in court opposite the women who had been attacked, his real name was known and was headed for newspapers throughout the northeast. He was neither "Jonathan Hall" nor "Smith." He was not thirty-five but just thirty-three. And he was not from South Carolina. He was a respected family man from Hartford, Connecticut. If the women in court took advantage of the opportunity to look at Horace Wells closely in the light, they would have seen a lanky man with sharply drawn features and thinning auburn hair. Pale under the best of circumstances, he was sickly on the day he appeared in court. Though normally fastidious, he was also badly in need of a shave. People who knew him might not even have recognized him, but then no one in New York City did know Horace Wells.

About a week before, on January 15, Wells had moved from the home he shared in Hartford with his wife and young son to a set of rooms on Chambers Street in Manhattan. He attended to business without delay, placing several newspaper advertisements on January 17, including the one that ran in the *Herald*:

> H. Wells, Surgeon Dentist, who is known as the discoverer of the wonderful effect of ether and various stimulating

> gases in annulling pain, would inform the citizens of New York, that he has removed to this city, and will for the present attend personally to those who may require his professional services. It is now over three years since he first made this valuable discovery, and from that time to the present, not one of his numerous patients has experienced the slightest ill effects from it; the sensation is highly pleasurable.

To describe a medical procedure of any kind as "highly pleasurable" is a rather gratuitous thing to do. Or perhaps to describe any highly pleasurable sensation as a "valuable discovery," is more suspect. Either way, there is an absurdity about Wells's advertisement, born of his tortuous perception that a painkiller is necessarily a pleasure giver. In Wells's case, though, it was reckoned to be just that: During his stay in New York, he freely indulged in the inhalation of chloroform, one of the newer painkillers. Whether or not he was already in that habit when he left Hartford, he admitted to using it constantly during his first week in the city.

Probably Wells did start out by experimenting with chloroform in a scientific way, as he later claimed. If he had been trying to present himself to New York as an authority on painkillers, he would have needed expertise with chloroform. And in the mid–nineteenth century, the accepted research practice was the same as it had been for centuries: self-experimentation. In Wells's case it turned quickly into an addiction.

One of the earliest authorities on drug abuse, Dr. Norman Kerr, described chloroform addiction in 1888, forty years after Wells's experience: "Chloroform is speedier in operation than any of the other forms of inebriety except ether," Kerr wrote.

> Unless the mania be resisted and the disease cured, the inevitable consummation by death approaches with startling swiftness. Interspersed with the most transient visions of delight, the life of the chloroform inebriate is

> but a protracted misery. The visions in the early stages of the diseased manifestations are most agreeable, but later on they become weird and horrible.

So it was that Horace Wells described the use of chloroform as a "highly pleasurable" sensation upon his arrival in New York. He had moved to the city for a new start professionally, and surely with some hope of escaping the disappointments that had carved themselves into his life over the previous three years. Chloroform was his answer for both.

The rooms Wells rented, at 120 Chambers Street, were situated at the very hub of New York. New retail buildings were going up throughout that neighborhood in 1848, replacing dilapidated old boardinghouses. Across the street from Wells's rooms, at the corner of Chambers Street and Broadway, the country's largest and most famous department store, A. T. Stewart, was constructing a new building, the Marble Dry-Goods Palace: "one of the 'wonders' of the Western World," according to an uptown neighbor, the diarist Philip Hone. Broadway near Chambers may have been preoccupied with its new retail shops, but it also continued to house most of the city's theaters. In addition it boasted the best hotels in New York: the City Hotel and the Astor House. The prostitutes who paraded Broadway after dark lived in boardinghouses on the side streets: Reade and Duane Streets, parallel to Chambers to the north; Church and West Broadway, just to the west of Wells's address. At night prostitutes catered to customers who were their equal in "abandoned character," and drew them into the neighborhood.

By night or day, though, busy midtown was Horace Wells's new neighborhood and it was the people there who kept him company in New York. He spent Tuesday, January 18, the night of the initial attacks on the prostitutes, with one of his new friends, a person much rougher than anyone he would have known in Hartford.

"A young man with whom I had recently formed an acquaintance," Wells wrote,

> went with me to my office in Chambers St., and while there, he said a woman of bad character had spoiled a garment for him while walking in the street, by throwing something like vitriol upon him; that he knew who it was and would pay her back in the same coin. As I had some sulphuric acid in my office, which I was using in some chemical experiments, he requested the liberty of taking some of it, for this purpose. He accordingly cut a groove in the cork of a phial, so that a small quantity only might escape when it was suddenly thrust forward. He then said that he might get it upon his own clothes. I told him that I had an old cloak, which could not be much injured by the acid, as it was good for nothing—By his request I walked into the street with him, he wearing my old cloak, and I having on my ordinary over-coat. We proceeded up Broadway, and when about opposite the theater, he said that he saw the girl he was in pursuit of, and he soon gave her shawl a sprinkling; we then turned down Broadway, when my friend proposed to sprinkle some of the other girls.

That was going much too far for Horace Wells, gentleman of Hartford.

"I immediately objected," Wells continued,

> and told him that what he had already done was not in accordance with my own feelings, although it was done in revenge; and when we arrived at Chambers St., I took my phial and cloak, at the same time, two of his friends came up and I left him, supposing that I had dissuaded him from doing the mischief he proposed, which is as foreign to my nature as light is opposed to darkness. I then regretted exceedingly that I had countenanced in any manner the first act. On getting home I found that my cloak had apparently received the principal part of the acid which had escaped from the phial as the wind was blowing towards us when the act was done. On meeting with my

> acquaintance the next day, he said that himself and his two friends, whom I met the previous evening, had resolved to drive all the bad girls out of Broadway by sprinkling them with acid. I in vain reasoned with him against committing so much injury when he had not been harmed.

On that day and on Thursday, however, Wells forgot all about the attack, leaving the vial on the mantel in his rooms. No one seems to have requested his professional services, and that is just as well because anyone who did call on him would have found him, as he described it, "in the constant practice of inhaling chloroform for the exhilarating effect produced by it."

Friday brought a beautiful January day to New York: "another delightful spring-like day," according to the *Herald,* "and all nature seemed to imbibe new life." Except Horace Wells: He spent Friday lying unconscious in his rooms, the chloroform inhaler still hanging from his mouth. He lost track of time, and of the quantity of fumes he was inhaling. Outside, construction workers were making progress building the Marble Dry-Goods Palace; around the corner, the "Model Artists" were at the Broadway Odeon Theater, preparing for a final weekend of shows, in which they portrayed or even acted out the subjects of great paintings. Horace Wells was up in his rooms, suspended in oblivion.

Awakening in the evening—a temperate moonlit evening that felt more like April than January—Wells was exhilarated beyond anything he had ever experienced before. He spotted the vial of vitriol on the mantel, where he had set it the previous Tuesday.

"In my delirium," he wrote, "I seized it and rushed into the street and threw it at two females. I may have thrust it at others, but I have no recollection further than this."

Wells didn't come to his senses until after he was arrested.

The day after the arrest, on Saturday, Judge Osborne heard the testimony of Wells's accusers, but the specter of further jail time inside the Tombs didn't disturb Wells nearly as much as the thought of what was happening outside its walls. "My real name is now before the public as a miscreant, guilty of a most despicable act," he wrote.

As Horace Wells sat in court, listening to descriptions of his atrocities, his friends and relatives were in Hartford. His wife even wrote him a letter on that day, idly inquiring as to his activities. In fact, word of the arrest would not reach Hartford until Monday afternoon, two days—and an eternity—for a man such as Horace Wells to be alone, condemned with his own thoughts.

After the arraignment Saturday morning, as Wells prepared to spend the rest of the weekend behind bars, he received permission to return home to 120 Chambers Street in order to collect a few toiletries. Choosing a moment when he wasn't being watched by the officer who accompanied him, he slipped a bottle of chloroform into a bag with his razor and other belongings. Then Wells went back to jail again, for the second time that weekend, but for the first time knowing who he was and where he was going.

"Dr. Wells entered our prison late on Saturday evening," affirmed the Tombs prison physician, John C. Covel,

> and was placed in a cell as comfortable and convenient as any we have. About 10 o'clock next morning (Sabbath), I called at his room and I immediately entered into friendly conversation, in which I begged him to unbosom his mind to me, and if there was anything that I could do, I should be most happy to serve him—I saw that he was furnished with sufficient stationery for any purpose that he might desire. Nor did I discover at that time the slightest indications demanding medical attention. He conversed freely of his friends and his intention of writing to them to obtain bail, which I encouraged, and farther intimated that by their assistance he should regain his liberty in a few hours.

After Dr. Covel left, Wells attended religious services in the chapel, where his mood apparently changed for the worse. The minister chose in his sermon to speak on the negative effect of an early association with disreputable women. Wells appeared to be engrossed in the message and he emerged from the chapel with a dark cast in his eyes. Returning to his cell, he requested a candle and began to write a letter to a New York weekly, the *Journal of Commerce,* delineating his actions. After describing what he did during the previous week, he dwelled in anguish on what he *didn't* do, seeming to have been prodded by the topic of the day's sermon:

> One of those abandoned females who were examined yesterday [in court], stated that I had often addressed her in Broadway. Now I do most solemnly assert that the statement of the girl is utterly false; I never have, on any occasion, had anything to say to these miserable creatures. If myself alone was the only one to suffer by all the false statements, which may be or have been made respecting me, it would be nothing compared to the injury to my dear-dear wife and child. Oh! may God protect them! I cannot proceed. My hand is too unsteady, and my whole frame is convulsed in agony. My brain is on fire.

By seven o'clock that night Wells had calmed down, but his writing again tore through to the source of his agonies. "What misery I shall bring upon all my near relatives, and what still more distresses me is the fact that my name is familiar to the whole scientific world, as being connected with an important discovery: and now, while I am scarcely able to hold my pen, I must bid all farewell! May God forgive me!"

When Wells finished writing his letter to the *Journal of Commerce,* along with short notes to his wife and others, he set his gold watch out neatly on the table in his cell. He fashioned a knife out of his razor, fitting a piece of wood along the back edge and fixing it with wire and thread that he'd pulled from his bedding. Putting out the candle, he sat down on the bed, administered chloroform to himself for the last time, and used the razor to slice through a main artery in his thigh.

The next morning between eight and nine o'clock, a deputy jailer opened the cell door and found Horace Wells dead, sitting up on his straw bunk, with his head leaning against the corner of the cell walls. His face wore a grotesque mask: He'd doused a new silk handkerchief with chloroform, wedged it into his mouth, and secured it with another new handkerchief, tied around his head. His hat was on his head. Wells's right leg was hanging over the side of the bed, while his left was lying straight out on the mattress, cut with a six-inch gash open almost to the bone. The empty bottle of chloroform lay between his legs.

Reactions to Wells's death were varied. Dr. Covel saw the body within hours of its discovery and presaged the verdict of the official inquest: "suicide while under temporary insanity," he called it and so would the court. In New York the event was more important than the man: The only person to comment on the suicide in the newspapers used it as an excuse to accuse Police Department doctors of neglecting their duties. Another person who made a comment on the suicide, in a tragic way, was a student from South Carolina by the name of Murray who copied Wells's method in every detail and was himself found dead leaning against the wall of his bedroom in New York.

In Boston, Horace Wells was known as one of three rivals for credit in the discovery of anesthetics: Since the other two resided in Boston, Horace Wells was very tepidly mourned in that city. His memory was duly honored in Hartford, but perhaps that memory was of a younger Horace Wells, as he had been before the great discovery. The neighboring city of New Haven seemed to see him as he was at the end.

"He spent some days in this city last summer—" reported an obituary in the New Haven *Journal*,

> called at our offices—and we were struck with the intellectual merit as well as the modesty of the man. There was something peculiar in him. He remarked to us the extreme pain he suffered from the course of some medical gentlemen in reference to his discovery, and we

> formed the opinion that he was subject to great mental depression, amounting almost to disease—a fact his friends say was true of him.

Horace Wells had one other mourner worth mentioning: a mourner of a sort. The day after the suicide, a police officer in New York took enough interest in the case to visit the hospital and see the prostitute whose face had been maimed by vitriol. The doctors told him there was no such case and never had been. The officer went back to the Broadway neighborhood where the attacks had taken place and called on the woman who had testified about the maiming. She told him she didn't know anything about it and shut the door.

The last moments of Horace Wells's life, then, were no different than any in the whole last year of his life: all that there were, were things that people said, though they didn't really know.

The man who died at the Tombs with a silk scarf stuffed into his mouth and a hat on his head, staring straight into the air, was a chloroform addict. But in Connecticut, Wells was a great man. To apologize for the confusion between the two, he proved himself willing to die.

16

IN THE LOBBY OF THE WILLARD HOTEL

On January 26, 1848, while newspapers throughout the nation were still reporting on the lonely death of Horace Wells in New York two days before, the trustees of Massachusetts General Hospital released their report on the discovery of etherization. It was considered newsworthy as an official judgment on the controversy. Terming Wells's claims "unfounded," the report even questioned the wisdom of the State of Connecticut in sanctioning his claim of discovery. According to the hospital Wells had not done anything that Humphry Davy did not do almost fifty years earlier, when he noticed the ability of nitrous oxide to cause insensibility.

As for Charles Jackson, the hospital was slightly more respectful, but found that he had neither discovered any new properties of sulfuric ether nor, damningly enough, ever tried even one experiment with the compound "on man or animal."

"Upon the whole, then, it seems clear that to Dr. Morton the world is indebted for this discovery," concluded the report, acknowledging that, "but for scientific knowledge and sound advice, Dr. Morton would not have made it at that precise time, and might have failed to do so at any time."

Mass General exhibited more than just moral support for Morton, taking up a collection among the doctors in May 1848. Morton received the proceeds, one thousand dollars, in a silver box with an inscription that read: "For William Thomas Green Morton, who has become poor in a cause which has made the world his debtor." The largess of the hospital staff was all the more remarkable considering that by 1848, hardly anyone in Boston's medical community could stand the sight of William Morton. Even the men who had once befriended him were wincing at the mention of his name. Dr. Augustus Gould, the most genial of the lot, was heard to mutter: "The subject of the discovery of etherization has become so offensive to us all in this vicinity, that I would gladly avoid ever alluding to it again in any shape."

Neither William Morton nor Charles Jackson grew in any way through their proximity to the discovery of anesthesia. Both tried to use it only in order to rewrite their own pasts, to correct the old wrongs they both dwelled on for years. For Morton: that his parents had gone bankrupt and the family's hopes had devolved on him long before he was prepared. For Jackson: that he had been pressured to return to the United States from France, where he had been a part of a world he respected; and even more pointedly that he had been robbed of the discovery of the telegraph. Those were the wrongs that the discovery existed to set right in the lives of Morton and Jackson—according to Morton and Jackson.

In early 1848 Morton was still awaiting the verdict of the French Academy of Sciences. In the autumn of 1847, he had submitted his version of evidence to the French Academy, under the title *A Memoir to the Academy of Sciences at Paris on a New Use of Sulphuric Ether.* In it he claimed that he had thought of the idea of using sulfuric ether in operations during conversations in late 1844 with Charles Jackson, then his part-time tutor. While Jackson never mentioned Morton by name in *his* letter to the academy, Morton graciously wove Jackson's name throughout his recollection of the incidents leading to Ether Day. However, Morton repeatedly insisted that he, not Jackson, connected the basic properties of sulfuric ether to a possible use in surgery.

By the time Morton submitted his *Memoir,* he was no

longer a businessman; he was a beggar. The last paragraph of the "Memoir" enlists Morton's new tone and covers the points that he would beat like a drum for the rest of his life:

"In justice to myself," he wrote,

> I should say that I took out my patent early, before I realized how extensively useful the discovery would be, and beside the motive of profit and remuneration to myself, I was advised that it would be well to restrain so powerful an agent, which might be employed for the most nefarious purposes. I gave free rights to all charitable institutions, and offered to sell the right to surgeons and physicians for a very small price, such as no one could object to paying, and reasonably to dentists. I had little doubt that the proper authorities would take it out of private hands, if the public good required it, making the discoverer, who had risked reputation, and sacrificed time and money, such a compensation as justice required. But as the use has now become general and almost necessary, I have long since abandoned the sale rights, and the public use the ether freely, and I believe I am the only person in the world to whom this discovery has, so far, been a pecuniary loss.

In March 1848 *Littell's Living Age*, a national magazine of news and culture, devoted an extensive article—forty-two pages of tight text (and no pictures)—to the ether controversy. The piece was contributed by a well-known lawyer named Richard H. Dana, Jr., and it was certainly a well-written article, Dana having been the author in 1840 of a bestselling book that is still considered a good tale today: *Two Years Before the Mast.* The son of a poet and editor, Dana had quit his studies at Harvard in order to ship out on a sea voyage to California. His book, the result of firsthand knowledge of the brutality dealt out to common sailors in the 1830s, led directly to reform and was an early model for both sea stories on the one hand, and the American liberal canon on the other. The article that Dana wrote for the

Living Age was well supported, but it was openly favorable toward William Morton.

Another man of letters, Edward Everett, seemed to agree with Dana's conclusions, and spoke for many people when he observed to an old friend:

> It cannot be denied that Dr. Jackson was himself principally to blame if he lost a part of the credit to which the Discoverer was entitled. He seems greatly to have neglected the idea when it had once occurred to him and to have taken much less pains than might have been expected to establish the existence and diffuse the knowledge of this wonderful agency.

To all such criticism Jackson would reply that he was too busy to devote himself full-time to the defense of an idea—and his abiding implication was that a man such as he should not have to campaign for credit. Whatever his perspective on the ether controversy, however, he was indeed right in saying that he was a busy man. Charles Jackson was a man of many pursuits, and many controversies. Nonetheless he felt compelled to respond to the Dana piece.

Three months later, in June, the *Living Age* bulged with another article on the controversy, a thirty-two-pager written by Charles Jackson's lawyers, Henry and Joseph Lord. The editors, for their part, warned the world in a preamble that "no party can have any equitable claim upon us for another long article on the subject." However, they admitted to having been swayed by Jackson's version of events, "which give the honor of the great discovery to a gentleman whose studies and habits are such as to make his claims probable."

"Of Mr. Morton we have learned more than we shall say at present," the *Living Age* editors noted tartly.

What the editors had undoubtedly learned about Morton was that he had been a crook out west. After conducting a relentless investigation of William Morton, Jackson's lawyers had compiled a dossier thick with affidavits and articles from Rochester, Cincinnati, St. Louis, and Baltimore, detailing Morton's schemes as a con man. The dossier was Jackson's secret weapon, and he was waiting for the right moment in which to spring it. With it, he would show that there was a pattern to Morton's life, a way of perceiving and then pillaging the world, whatever the circumstance.

Charles Jackson was perfectly right that Morton would never change, that his scurrilous past was his only future. Jackson would never change, either, though. In 1849 he was recalled from Lake Superior, under pressure to resign as a U.S. geologist.

The fracas began when Jackson's two lieutenants complained to the secretary of the interior about his management of the surveying project. In response Jackson did what he always did: He induced his friends among the elite of the nation's scientists to write letters on his behalf. He claimed that he was a victim of political intrigue, an Eastern conservative (or "Whig," as he called himself) crushed in the path of Western liberals. And then he called in his brother-in-law Ralph Waldo Emerson.

In May, Emerson wrote to his own brother William, "My time was sadly occupied for a fortnight in Dr. Jackson's affair. We now get from him assurances that he is to be left undisturbed in his survey." By June, however, the tide was against Jackson, and he submitted his resignation. His protestations bore a familiar refrain, as he complained that the officials in Washington, "had other ends in view than those of science or public service and utterly disregarded every document that went in my favour."

That summer of 1849, William Morton was making better progress in Washington than was Jackson. Mass General sent a petition supporting his right to recognition to Congress, where yet another select committee of the House of Representatives was meeting to investigate the discovery of etherization. In the midst of the debate Morton received an impressive endorsement from Daniel Webster, one of the most esteemed men in the Senate, or indeed in the nation.

In July the select committee came to a decision.

On July 22, 1849, William Morton sent a telegram to Nathaniel Bowditch, the Massachusetts General Hospital trustee who was his most spirited defender. "The Committee have awarded the credit of the discovery to me," it read. That telegram may represent the very highpoint of Morton's long Washington campaign, as the representatives most closely charged with the decision about the one-hundred-thousand-dollar award came to his side. The money seemed to be promised, and with it, a life of ease and grace.

Washington was not nearly so simple as that, however: not then, not ever. The petition for an award stalled before reaching the full House in the form of a bill. Even so, a coalition for Morton did seem to be building in the autumn of 1849. Then, at the end of that year, a doctor from Georgia named Crawford Long wrote an article for the *Southern Medical Journal*, claiming that he had administered sulfuric ether during minor operations as early as 1842—four years before Ether Day.

According to the article Dr. Long had started a flourishing fad for ether frolics in the neighborhood of his practice in about 1840, after telling a group of young sports who requested a bag of nitrous oxide that he could not supply them with that gas, but knew of something just as good. Dr. Long himself would often indulge in a whiff of ether. A serious country doctor with a family, he did not consider himself degenerated in any way by his ether whimsy, but rather amused. If also bruised: He often came to with hurts and scrapes he had no memory of receiving.

On March 30, 1842, Dr. Long was consulted by a fellow ether user named James Venable, who wanted to have two small tumors removed from the back of his neck. For some time, however, Venable had delayed the procedure from, as Long put it, "dread of pain."

"At length," Dr. Long wrote in his article, "I mentioned to him the fact of my receiving bruises while under the influence of the vapour of ether, without suffering, and as I knew him to be fond of, and accustomed to inhale ether, I suggested to him the probability that the operations might be performed without pain." Venable gladly submitted to the impromptu ether frolic, and the operation proceeded without any infliction of pain whatsoever, so the patient reported afterward. Two months later the same patient underwent a second operation with good results.

In July, Long administered ether to an African-American boy who had a toe amputated under its influence. That was the extent of Long's initial experimentation, two patients and three etherizations: His practice did not present any further opportunities for trial. However, all the indications were positive.

"The question will no doubt occur, . . . " wrote Dr. Long in the article. It is hardly necessary to finish the quotation. The question definitely occurred: Why didn't he spread the news of his great discovery? He gave two reasons. First, he needed more experiments in order to prove that insensibility had been induced by the ether and not by some suggestion of the imagination. Second, in related reasoning, he did not want etherization to be confused with mesmerism. He needed a more serious operation in which to try it, and his practice rarely offered anything more serious than minor removals and the amputations of fingers and toes.

Crawford Long was not at such a great disadvantage, however. Had he felt any urgency regarding his discovery, he could have recommended it for further experimentation to surgeons with more active practices, even in the South. Long was an 1839 graduate of Jefferson Medical College at the University of Pennsylvania—which was then the premier medical school in the country. He would have been within his province to contact the teaching hospital there with his discovery. He did nothing, though, and meanwhile, suffering during surgical operations was prolonged for another four and a half years.

Long's story reinforced a philosophical question that was essential to the ongoing ether controversy—whether the greater merit rested with the *discovery* or the *introduction* of anesthesia. If it was the discovery, even in isolation, then Crawford Long was the hero. If it was the introduction, then the combination of Wells, Morton, and Jackson held the fore.

If—to continue the progression of thought that followed Long's proclamation—if it was a case of sheer discovery, no matter the follow-up, then Long himself had competition from other jump-up behinders. One story that surfaced concerned a Rochester doctor named William E. Clarke, who administered ether to a patient known only as Miss Hobby. While she was under its influence, a dentist extracted one of her teeth. The date of that etherization was January 1842, two months in advance of Crawford Long's first experiment with ether. The story of Clarke's use of ether was published only later, and remained unsubstantiated. Though William E. Clarke went on to become a well-known practitioner in Chicago, he never commented on the etherization story, as far as is known. (It is even possible that the story came from a man who just happened to be in Rochester in January 1842, William T. G. Morton.)

After an initial flurry of excitement over Crawford Long's article, his cause sank quite neglected under the claims from New England. Long, who was living in the quaint university town of Athens, Georgia, at the time he wrote his article, was only blandly noted as a claimant in France. One medical writer there referred to him as the "Greek physician," which may indicate the level of investigation exhausted in Paris on Crawford Long (of Athens).

With help from Georgia's congressional delegation and the innovative southern doctor, J. Marion Sims,* Dr. Long's claim would eventually gather strength in Washington. Anyone entering the pursuit of congressional recompense, however, would have to match William Morton, who was by 1850 a full-time "lobbyist," a term used even then to describe him. By that time, Horace Wells had better representation than he had ever had

*A founder of the practice of gynecology.

while he was alive, in the form of Connecticut senator Truman Smith. Seeking an award for Wells's impoverished widow and son, Smith worked zealously both in and out of Congress, even writing a book on Wells's plight. As for Charles Jackson, his campaign was purely defensive, as far as congressional action was concerned; his only goal was to keep the Morton and Wells camps from receiving any money or official recognition.

The recognition that Jackson coveted would come, if anywhere, from Paris. In 1850 his waiting was over; the Academy of Sciences presented its decision on the ether controversy. On the one hand, he had confidence that he would win its approval. On the other, he dreaded the thought that William Morton might win it instead. However, one possible verdict was even more repugnant than that . . . .

The announcement from Paris specified that Jackson and Morton would share in the honor. Each would receive a gold medal representing the Prix Montyon, as well as a cash prize. "The Commission," read the official report,

> proposes to the Academy, that a prize of 2,500 francs be awarded to Mr. Jackson for his observations and experiments on the anæsthetical effects produced by the inhalation of Ether, and a similar prize of 2,500 francs to Mr. Morton for having introduced this method in the practice of surgery, after the indications of Mr. Jackson.

The compromise suited people around the world who could easily accept that the magnificent discovery had had two authors. In the United States, however, the French decision served only to perpetuate the controversy. William Morton the winner would have meant Jackson the vanquished, and likewise the other way around. For two men who could not abide a tie, the decision was only an incitement to fight on. And as long as the United States had a Congress, William Morton had a battleground.

In 1852 the 32nd Congress received a new petition specifying a formal expenditure of $100,000 in lieu of moneys lost by

William Morton from the government's failure to uphold his patent.

"The discovery of Etherization," reported the pro-Jackson *National Police Gazette,* "has recently been brought for a second time before Congress, on the petition of a (so-called) dentist, William T.G. Morton."

"The ether controversy is coming to a crisis," wrote the Washington correspondent for the *New York Times,* March 19, 1852, after a House select committee issued a report voting unanimously in favor of Morton. By late summer the allocation of $100,000 to Morton was attached as an amendment to the Army Appropriations Bill. A Senate select committee made a full investigation during the autumn, calling witnesses by the score. First the committee tried to establish whether or not there was a precedent for making an award to an inventor in lieu of patent rights. And so a list was drawn up of the patents previously purchased by the government. Most of them were strictly military in nature. The award most similar to that sought by William Morton was $76,300, which went to the heirs of Robert Fulton, "for the benefits conferred upon the country by his improvements in navigation by steam." Many senators nonetheless remained troubled. They doubted that the Republic should bestow largesse on citizens, as though from some shadow of a royal prerogative. Others flatly denied that sulfuric ether, a common compound, could be patented—and if the patent were worthless, then the government had no reason to pay $100,000 for it.

When the debate moved forward, it addressed the dreaded question of who deserved credit for the discovery; Morton, Jackson, or Wells. Morton, on the scene in Washington during most of the hearings, oversaw an effective case, in which former colleagues helped show that he *might well* have originated the idea for anesthesia himself, with slight assistance from Horace Wells and Charles Jackson: Testimony by surgeons from the staff of Mass. General proved unequivocally that he was the first to make a public demonstration of ether anesthesia.

Dr. Jackson initially replied to the select committee's request for evidence by denying that there was anything for the committee to investigate: that the matter was *res judicata*—already settled. The Academy of Sciences in France had deemed him the originator, which was to say, the true discoverer. However, Jackson sent a legal staff, in the form of one young lawyer, to represent his interests and cross-examine witnesses brought forward by Morton's side.

As the congressional investigation slogged along, Dr. Jackson bided his time and readied his secret weapon, the dossier on Morton's activities as a con man. "Things at Washington in Status Quo," he wrote to Ralph Waldo Emerson in July. "All ready to fire on the Enemy the moment he comes in range."

"Morton still pouring out thirst for members of congress daily," he added, referring to Morton's overt hospitality in Washington.

Many people charged that Morton entertained Washington officials royally in the suite he took at the brand-new, ultra-swank Willard Hotel—attempting to influence members by "the lavish use of champagne, cigars and oyster-suppers," in the words of one reporter. It was an expensive effort, but Morton considered it vital to mix with the right people and win their confidence. That had been, after all, his early training.

In autumn Jackson decided that the time was right to deliver the shocking dossier on Morton's past to his rival's new friends in the Congress. The bombshell, however, failed to explode. As soon as the statements were described to the committee, Morton himself responded by declaring his *insistence* that they be introduced into evidence, so that he could not only rebut them, point by point (and city by city), but introduce a mass of proof he said he possessed illustrating Jackson's habitual mendacity. The counterattack seemed to terrify the congressmen with the specter of refereeing yet a new wrestling match between Morton and Jackson. The committee flatly refused to consider Jackson's thick file on Morton, terming it "evidence of general character . . . not relevant to the issue." And so the story of Morton's career as a swindler never became public.

In the end the select committee found that William Morton

deserved the credit for the discovery and the award of $100,000. The full Senate, however, refused to address the issue as an amendment. That killed Morton's chances for another year.

The following session, in spring of 1853, Senator Truman Smith of Connecticut introduced a bill on behalf of Horace Wells's survivors. It, too, received the attentions of a select committee, which moved to send the decision to the judiciary, in the form of the Circuit Court of Northern New York, that being considered a relatively convenient but neutral battleground. To try the issue in court might have produced some clear-cut decision, but the House failed to follow the Senate's vote in favor of the plan and so the issue was sent back to the only tribunal it ever really received in the United States: the public forum.

To rejoin the battle in that forum, Morton published the high points of his recent season in the halls of Congress in a book called *Statements supported by evidence of W.T.G. Morton, M.D., on his claim to the discovery of anæsthetic properties of ether, submitted to the honorable select committee appointed by the Senate of the United States, 32nd Congress, 2nd session, January 21, 1853*. The appendage "M.D." after Morton's name was taken from an honorary degree he'd received from a small medical school in Maryland. The book was meant to look like a government document, and was taken to be one by most people who read it. However, it left out most testimony unfavorable to Morton and included "reports" that had never been issued by the committee. One of these, for example, was labeled "In the Senate of the United Senate." He meant the United States.

In the 1850s, when William Morton was spending his winters in Washington, he was spending summers at Etherton Cottage in Wellesley with his wife and their young family, five children having been born between 1845 and 1857. In addition Morton set up his mother and father in a small house nearby. Etherton Cottage was a renowned operation in its day, a model farm with its own hydraulic pumping station and steam engine for milling and sawing. Morton's specialty was breeding, a pursuit for

William and Elizabeth Morton with their children at Etherton Cottage, Wellesley, Massachusetts, c. 1860.
(Van Pelt Library, University of Pennsylvania)

which he won prizes at agricultural fairs in Massachusetts. In fact, even when William Morton was in Washington, he kept abreast of every development on the farm. He and his farm manager would stand in telegraph offices in Washington and Wellesley, respectively, shooting messages back and forth by the hour, as though in conversation. It was akin to an early form of E-mail: electromagnetic-telegraph mail. However, it was an expensive way to chat about prize hogs and melons in the 1850s. It was an expensive way to chat about anything.

In 1853 Morton's famous friend Mrs. Sarah Josepha Hale wrote about Etherton Cottage for her magazine, *Godey's Lady's Book.* She had never visited it herself but quoted someone else as calling it, "a picturesque building of the English style of rural architecture. The prospect from its every window is, of course, superb. In the foreground are the serpentine walks, rustic summer-houses, flower-beds, young trees, sparkling streams and other appurtenances of the mansion itself."

The lordly style in which Morton lived was apparently better appreciated from afar, however. His neighbors in Wellesley mocked him and teased the children, who were tutored in their own schoolhouse on the grounds. For a while the expenses of maintaining the Morton estate were covered, or at least defrayed, by a small factory that Morton operated, making artificial teeth. His campaign for the grand reward from Congress proved a distraction, though, and the factory was closed by the early 1850s. Every year or so the doctors of some city would take up a collection for Morton and that helped him to run his household. He borrowed money copiously and repaid it rarely; that also helped. In 1853, however, William Morton was even more in arrears than usual, and in even more desperate need of a new source of money than usual. He felt that previous efforts in Congress had softened the legislators and that to take advantage of it, he needed to make a greater effort. (A greater effort, according to many reports, meant even stickier bribes to even more congressmen.) With Truman Smith retired and out of the way, the time was at hand to collect the one hundred thousand dollars he felt was his due. That was Morton's reasoning.

And then he met William S. Tuckerman. A young man of strong ambition, Tuckerman held an enviable post as the treasurer of the most important railroad in the Boston area, the Eastern Railroad. Founded in 1836, the Eastern carried just short of a million passengers in 1854 and earned a healthy dividend for its stockholders. In the company as it was organized at the time, the treasurer was the active manager of the line. He was also the trustee of its money.

17

EMERSON'S WIFE'S BROTHER

When the celebrated Louis Agassiz arrived in Boston in late fall of 1846 to deliver a course of lectures in Cambridge, Charles Jackson, his fellow geologist, was one of the first people he met. Agassiz, a Swiss by birth, was an astute man who had attained his own measure of glory in the scientific circles of France and Germany; he was intimately aware of the subtleties on which Charles Jackson's claim in the ether controversy rested. He was also well-acquainted with the distinction Jackson made between the originator of a discovery (that is, Jackson) and its comparatively inconsequential demonstrator (Morton). However, Agassiz suggested an even more exacting test, involving not merely the inspiration behind the discovery, but the intent:

"I remember distinctly," Agassiz wrote in 1852, referring to an encounter he had had with Jackson after a professional meeting in October 1846, "walking home with him and several other gentlemen, among whom I remember Dr. A. A. Gould of Boston, and discussing the extent of Dr. Jackson's claim in this matter.

"I asked Dr. Jackson whether, in case Dr. Morton had killed the first patient to whom he applied ether at Jackson's suggestion, he, Jackson, would have claimed the whole merit for the discovery, or even the credit of the suggestion? To which he answered nothing," Agassiz said.

For Agassiz, Jackson's silence was damning, yet that was not quite how Dr. Jackson recalled the same conversation.

Ralph Waldo Emerson, jockeying to gain Agassiz's support in the controversy, consented in 1859 to ask his brother-in-law Charles Jackson a series of questions, as supplied by Agassiz. The first ones sought to define the nature of the discovery as specifically as possible, but the last concerned the responsibility for the success or failure of the operation. Emerson was able to report back to Agassiz that Dr. Jackson had definitely taken that ultimate responsibility, and before witnesses, well in advance of Ether Day.

"To this point, by the way," Emerson told Agassiz, "the doctor remembers with pleasure and has before cited it to me, a remark once made by you in a conversation with him and Dr. Gould, 'O, if Doctor Jackson took the responsibility for the advice, then the discovery is his.'"

To Jackson that statement had been a declaration: conclusive unto itself. To Agassiz it was nothing more than a proposition, one to which he never felt able to add a substantiated conclusion.

Another Harvard scientist, Prof. T. T. Bouvé, was troubled by a different ambiguity in Jackson's claim. Accepting at face value Jackson's insistence that he had originated the idea for etherization in 1842, Bouvé noted without satisfaction that that left a gap of four years before the discovery was made public. A gap such as that gave Jackson the same precarious claim to glory—or infamy—that belonged to Crawford Long in Georgia.

Bouvé supported his position—and described the impossibility of the controversy—with a recollection. He said he had taken

> the ground that the world was indebted to both Jackson and Morton for the great boon; to one as the scientific discoverer and suggester of its use in surgical operations, to the other for his application of it and its practical introduction. Dr. Jackson, learning of this, upon meeting me, remarked that I was thought not to be friendly to him in the matter.

Anyone who gave Morton even half credit gave Jackson none, in Jackson's view.

"I then said," Bouvé related, continuing his story, "'Doctor, you have known for a long period what Mr. Morton is now demonstrating to be true, but have allowed it to remain a dormant fact in your mind. If he had not sought information from you, might it not have remained so for years longer?'

"He answered that it might," Bouvé reported. To him the delay indicated unequivocally that Jackson was missing some quality necessary to bring the discovery to fruition. Morton, whatever else he lacked in the way of integrity or even intellect, possessed that particular quality.

Jackson's four-year gap (the delay from his purported origination of the idea in 1842 to Ether Day in 1846) and the November shift (from dissociation with the discovery to the attempted usurpation of it) lend an aura of mystery to his claim. Scientists, including Agassiz and Bouvé, who were familiar both with the processes of discovery and with Dr. Charles T. Jackson, doubted that he believed in the worth of etherization—until after Ether Day renewed a long-expired dream and made painless surgery possible. Those with the most respect for Jackson's fertile mind suspected that he was the sort to brush by a hundred superb ideas a day, like dandelions around his ankles, without stopping to appreciate the wonder of any single one of them.

Charles Jackson, his old friend C. A. Bartol once said, "was simple as a child and veracious like the sun." Those scientists who were disturbed by the weak spots in Jackson's claim presumed that his version of the story was at least truthful. In October, when Jackson was taking only a furtive part as a participant in the discovery, he was treating it as a business transaction, first billing Morton for five hundred dollars as a consulting fee and then, later, haggling over a royalty arrangement, and finally signing a patent application as coassignee. What could he have been thinking at that time? If he truly thought that the discovery were of international importance, he would have so apprised the Academy of Sciences in France. That is what he did in November, after the November shift: after he knew (secondhand from men such as Henry Bigelow, A. A. Gould and John C. Warren) that etherization was indeed a breakthrough.

In early October, when the possibilities of etherization were being proven, he sent his sister, Lidian Emerson, to N. C. Keep's office, where she underwent an extraction. Advised by her brother, "she did herself carry to Dr. Keep the exact details of the use of ether for use in her own case," Waldo Emerson recalled, "and the experiment proceeded well." It was the only ether experiment definitely instigated by Charles Jackson. After Lidian's successful experience, Jackson must have been convinced of the efficacy of etherization, and so he contracted for a percentage of the profits. In that era dentistry was a commercial, not a humanitarian, undertaking. One would not feel a moment's angst over profiting from the relief of pain in dentistry. And it appears that Jackson didn't.

Charles Jackson could not have been thinking of etherization in terms of surgical use before Ether Day. That would explain both the four-year gap and the November shift. Having noted that ether produced a state of insensibility effective enough for a brief dental procedure, he did not feel any humanitarian compulsion to develop the idea, and so four years passed, before William Morton the dentist came along, inquiring about just such a compound. As to the shift, his attitude would naturally change if the dental novelty he had been developing suddenly turned out to be a medical miracle. And that point would clarify yet another blur in Jackson's claim: why he applied for a U.S. patent. After his November shift, Jackson joined the rest of the medical community in denouncing the patent—and yet his name was on it. He explained that rather embarrassing inconsistency by saying that the patent process represented the only way he could establish a claim to the discovery in his own country.

That there is plentiful room for conjecture regarding Charles Jackson's actions implies, if nothing else, that there was something disjointed about his story. Agassiz found one trouble spot, Bouvé another, and there are others: In those blurry patches one can catch a glimpse of a man as "simple as a child and veracious like the sun," building up towers of blocks that kept falling awry.

In the months following Ether Day, Jackson did finally administer ether himself during operations in and around Boston. His roaming mind, however, soon left that to surgeons already devoted to improvements in surgery. He turned to yet another original use for etherization: in the treatment of the mentally ill. According to Dr. Luther of the McLean Insane Asylum, which was part of Mass General, Jackson often visited to administer ether to the insane, giving them, in many reported cases, temporary comfort. Before the end of the winter of 1847, however, Jackson returned to his full-time career as a geologist.

One aspect of Jackson's reputation was based on astonishment at his sheer energy: No amount of bother dissuaded him in the laboratory or in the field. "We might say he lacked patience," wrote Franklin Benjamin Sanborn, an author who'd come to know Jackson in Concord, "if he had not labored so unwearyingly in the most exacting of the physical sciences—where indications are nothing, and verifications infinitesimally difficult." In honor of his work in exploring the geology around Lake Superior, a colleague proposed that a certain mineral, identified as a hydrous prehnite, be named "Jacksonite." The honoree, however, tested the mineral on his own and found that it was not, in fact, hydrous (composed of water). Thus Jackson himself stripped "Jacksonite" of its name and sent it back to its gracious discoverers, just an anonymous bit of rubble.

"Adventure and novelty had great charms for him," noted Sanborn. That was a fact, but there was a qualification attached. There was always a qualification on statements about a man with so many hairpin turns in his personality. "Adventure and novelty had great charms for him," said Sanborn, "and he accepted ungraciously the wooden proprieties of Boston society, in which he was seldom quite at home."

In Paris, Charles Jackson had made himself amenable, and perhaps just a bit obsequious, before men he considered worthy. In Boston, however, Jackson was a maverick. For a pure scientist that is a fine role. For an ambitious one, however, it is foolhardy. "He had that disdain for trifles and trifling persons which is often the mark of a gallant and imperious spirit," was the way Sanborn put it.

"That monomaniac," was how many others in Boston regarded Charles Jackson. A monomaniac is a person obsessed with one thing, and in Jackson's case, that one thing was claiming credit for inventions: the electromagnetic telegraph on one hand, anesthetics on the other. This delightful/difficult man, with his amusing stories of the trail, drove potential allies away, giving them little choice except to regard him either as the greatest scientist of his day—or as a monomaniac.

Emerson, who stood by Jackson as no one else did, recognized that his brother-in-law had a prickly personality under the best of circumstances, and a volatile one in the midst of the ether controversy. At the end of his letter to Louis Agassiz, that deposition which so carefully reported Jackson's answers to the professor's questions, is a gentle apology for his brother-in-law's temperament on previous occasions. "What strikes me most formidably is the clear and assured tone of all Doctor J.'s conversations with me on this matter," Emerson wrote. "He has run the gauntlet so often and has come at last to see that there is a power in facts against all opposition, that he has become perfectly peaceable and amicable on the subject, and quite willing to say nothing and let his claim rest on the very testimony offered by his opponents.

"I am, however, weak enough to think it of high importance to him that you should be accurately informed of the facts he rests on," Emerson finished.

Emerson wasn't weak. The fact that he took the controversy so squarely on his own shoulders shows why he was adored in his family and beyond it. Moreover, it shows why Jackson's cause did not lag as long as Emerson was alive. Over the course of more than thirty years, he poured out letters to influential people, whether he knew them previously or not, arguing the unpopular controversy.

Privately, though, Emerson never quite understood Jackson's course in allowing the controversy to envelop and even destroy him. "The pertinacity of the opposition to Dr. J.'s claims with us does indeed astonish me," he wrote to his wife, Lidian, "and I can easily understand should dishearten him. But he who discovered so much, can discover a great deal more, which his swindlers and enviers can not. He ought to leave defending his rights . . . and forget it in his laboratory."

Charles Jackson had another champion in J. B. S. Jackson (no relation), one of the tutors who had prepared him for medical school many years before. James Jackson had cofounded Massachusetts General Hospital with John C. Warren in 1811, despite the fact that the two disagreed on practically everything. They found that out early enough. Recognizing the potential of a partnership between them, though, they made a pact not to squabble publicly or to let their basic antipathy hinder the progress of the hospital. Remarkably their pact held over the thirty-five years that the two worked together. On the ether controversy, Dr. Warren remained neutral, while J. B. S. Jackson sided with Charles Jackson. The hospital's Morton side was led by the dauntless, and very powerful, Nathaniel Bowditch.

"I am sorry that I cannot claim the unanimous support of the Massachusetts General Hospital in behalf of Dr. Jackson," Emerson wrote to a Congressman in the 1864 clash over the Morton claim:

> I believe there are personal differences even among the Doctors, and perhaps Dr. Jackson had the misfortune to offend some of his seniors. Dr. Warren, at its head, was always his firm friend. But Mr. Bowditch, one of the Trustees, a gentleman of great worth, and some heroic

> traits—in some feeling, I believe of arrogating the glory to the Hospital—set himself to destroy, if possible, Dr. Jackson's title to the discovery.

On another occasion Emerson was even more blunt: "The true account of Morton's degree of success with the public," he wrote, "lies in the intent which a few efficient and influential persons took in winning such a rare laurel for their Massachusetts Hospital." There were accusations from many other sources, too, notably Connecticut and Georgia, that William Morton had backed himself into a rather snug partnership with the Massachusetts General Hospital—and that that institution didn't care a whit about who actually deserved credit for the discovery of anesthetics. Morton happened to fit best into its own scheme for self-aggrandizement. That theory falters a bit when one considers that the hospital could have plucked a stranger off the street and stood a good chance of finding a better man to represent it.

William Morton was an embarrassment, first with his money-making schemes and then with his constant need for charity (in order to maintain his estate at Wellesley). His designation as the discoverer did succeed in focusing attention on Ether Day itself, which did occur in the operating theater of Mass General. That room, its walls having witnessed the very first public demonstration of anesthesia, was soon renamed the "Ether Room."

Had the hospital been as self-serving as some implied, it would have showered the glory of the discovery onto John C. Warren. He was the commanding general of the occasion: the one with the power to allow the experiment and the courage to actually go through with it. Ether Day was consistent with the rest of his career, casting American medicine as a leader in research and treatment, no mere follower of European advancements. Dr. John Warren never claimed the slightest credit for Ether Day, though many people at the time speculated that he could have, with his lofty reputation. Warren was, in fact, the one man connected with the discovery who set an example that could be followed—and which was followed, as American medicine came of age. He died in 1856 and can be regarded as a hero in the discovery of anesthetics, since he was alone among the powerful in never giving up hope.

The three sides in the controversy were drawn into a stalemate in 1854, when William Morton seemed closer than ever to his hundred-thousand-dollar reward in the halls of Congress. That is when Charles Jackson went roaming and very nearly managed to encircle the enemy.

"On March 8, 1854, early in the day, a stranger entered the old drug store in Athens and inquired for Dr. C. W. Long," wrote C. H. Andrews, who was then a clerk in the small pharmacy co-owned by Crawford Long in Athens, Georgia. "I told him Dr. Long was absent, but I thought he would be in in a short time and invited him to a seat at the fireside. In a few minutes Dr. Long came in and I said, 'This gentleman has called to see you.' The stranger presented his card introducing himself as Dr. Charles T. Jackson of Boston, Mass.

"Dr. Jackson was a 'spare made' man, angular, of five feet ten inches height, of swarthy complexion, with dark hair and eyes, and apparently forty years of age," Andrews continued. "After a few moments conversation, Dr. Jackson said to Dr. Long that he had called to see him for the purpose of comparing notes as to the 'first discovery of the anaesthetic effects of sulphuric ether.'" Dr. Long consented, calling upon young Andrews to serve as a witness to the ensuing display, on each side, of evidence and affidavits. "In their protracted conference," Andrews wrote, "they were frank, but slow, cautious and exact. It was a weary day's work." The next day Jackson left to complete the geological commission that had drawn him to the South, but within ten days he was back in Athens. For two days, he and Long remained cloistered in the office at the pharmacy, conferring on the ether discovery.

Andrews made his recollection in March 1900, forty-six years after Jackson's visit of March 1854. Despite the passage of time,

Dr. Charles T. Jackson, 1862.
(Francis A. Countway Library of Medicine, Harvard Medical Library)

his command of detail was good—although it might be noted that Charles Jackson's complexion was not naturally swarthy. He had been out in the field, performing his duties as a geologist, and probably developed a deep tan. And that Andrews regarded Crawford Long with "reverence and love," but that was not unusual; a great many people who knew the courtly doctor felt the same way about him. Andrews may have painted Long with a warm glow and Jackson in shadows, generally, but he stands as a witness to an otherwise obscure negotiation.

"Throughout their conferences," Andrews said of Jackson's second visit to Dr. Long's office,

> during the two days that Dr. Jackson was in Athens, to whatever proposition he made to Dr. Long for sharing the honor and benefits of the discovery Dr. Long replied: "My

claim to the discovery of the use of sulphuric ether as an anaesthetic rests upon the fact of my use of it on March 30, 1842, of which I have indisputable evidence under oath and from reputable citizens."

Jackson, who could not abide the implications of sharing credit with Morton or with Wells, called upon Crawford Long in order to strike a deal along just such lines with him. He not only respected Long's work, he appreciated the man's quiet demeanor. Most people who claimed to have discovered anesthesia were rather agitated, if not manic: Stated statistically, three-quarters of them were. Long was neither.

In the manner of agitated, manic men, Jackson was still talking about Morton when he praised the fact that Long was "not disposed to bring his claims before any but a medical or scientific tribunal." In other words, not before Congress.

The conference in Athens represented a way out of the controversy, at least for Jackson: a solution both scientific and politic. That solution did not transpire, however. Dr. Long wanted no association with another claimant. He presented his facts and stayed with them; no deals could change the fact that he had experimented with etherization in 1842 and deserved, in his mind, some accord in medical history. Certainly his integrity can't be questioned. Instead Charles Jackson left Dr. Long behind, and on the way home he was writing a long, detailed letter to Emerson all about Morton's lobbying in Washington—so wicked, it seemed to Jackson. And so bound to succeed, it seemed to him too. "Morton has men in both Senate and House committees who will do anything he wishes," Jackson wrote home, "and he is pouring out liquors like water to the members of Congress and Lobby men."

Morton continued to work the Washington trail: No matter how many times bills for remuneration were laid aside, he kept propping them up again somehow. In July 1858, Charles Jackson received a mysterious note from one of Horace Wells's relatives: his widow's brother Joseph Wales. "I am told that

Morton is again afoot and apparently preparing for another campaign at Washington," he wrote, making reference to the businessman William Tuckerman, speculating on a probable new strategy and the intervention of new allies.

"Be this so or not," Wales concluded, "we must extinguish Morton at Washington should he show himself there in any shape." Wales then asked Jackson to inform him of Morton's itinerary, if it was known to him. What can be gleaned from the note is that Horace Wells's side was amenable to Jackson's, and that both were arrayed against Morton. What cannot be gleaned is what Wales meant by "extinguish," a boldly threatening word, except that it passed between two men obviously exasperated by Morton's invincibility. Nothing came of the Wales threat, whatever it really was. Nothing could extinguish the greed of William Morton, a fact that may have had the greatest impact of all on the life of Charles Jackson.

18

MORTON'S CIVIL WAR

On the night before the directors' meeting of June 28, 1855, William Tuckerman paid a visit to one of the directors of the Eastern Railroad, Benjamin Reed. An older man, highly respected in business, Reed lived in a large house in a small town north of Boston. After an inconsequential visit of about a half hour, Reed accompanied Tuckerman out to the stable in back of the house. Reed was "distressed," as he noted later, being outdoors without his hat. But he was to become much more uncomfortable before he went into the house again. "I don't remember what his exact words were," Reed said of Tuckerman, "but the substance was, he was in trouble and wanted my advice; I asked him if he was a defaulter; he said he had loaned money and could not get it back in the time it would be wanted.

"He hesitated about giving me the names of the persons to whom the money had been loaned," Reed continued, speaking of Tuckerman's confession. "He said it would prevent him from getting the money which he was confident of getting." Tuckerman had used the railroad's money for secret speculations, including short-term loans.

The next week the annual stockholders' meeting of the Eastern Railroad had to be postponed due to the crisis. On July 30, however, three hundred stockholders packed a meeting hall to hear the worst possible news: Their treasurer was a thief. In 1853–54 Tuckerman embezzled more than $245,000, $30,000 more than the line's net income for the year. At that grim meeting the stockholders heard that Tuckerman's books "were kept in the most careless manner." He didn't bother with a cashbook but jotted entries on bits of paper strewn about. Those were the real accounts. The ones that Tuckerman was in the habit of showing the other officers in the rail line were beautifully kept and totally fictitious.

One stockholder got up to say that he had tried to warn the company's officers that there was something wrong with Tuckerman's accounts. Then another, a Mr. Hildreth, stood up to tell what he knew: that earlier that year, Dr. Morton, "of the Ether controversy" had drawn $20,000 on a note issued by W. S. Tuckerman.

The news rippled through the meeting, and shouts were heard for an official disclosure of Tuckerman's dealings with Morton. That was not likely to happen.

Reed later described his first conversation with Tuckerman regarding the identities of those who received the company's money in loans. "He named Dr. Morton, but begged me to keep the name secret, as it would prevent him from getting the money; I said I did not see how he could keep it secret." Ultimately, though, the company concurred with that strategy. Tuckerman had loaned Morton the money specifically for use in prosecuting his claim in Washington; according to the arrangement they had made, Tuckerman would receive a proportion of the award, supposedly $50,000, when it was in hand. The amount of Eastern Railroad money that Tuckerman invested in the scheme was never satisfactorily disclosed.

"When I asked if the amount was $5,000, he said it was more," said Reed. "I went on as high as $20,000, in each case before reaching that sum, when I asked him, he said it was more." Some estimates placed the amount at much more than that, but whatever it was, the company let Morton keep it, on the assumption that after he received an award from Congress, the company would attach it. In the meantime the rail line canceled dividends, scaled back train service, and further reduced expenses by firing about seventy employees. Among those cut was the president of the line. Tuckerman had resigned from the Eastern Railroad but faced charges for embezzlement. Defended by the celebrated lawyer Rufus Choate (who also advised William Morton in a continuing capacity), he actually avoided a sentence. Within a year, however, he was arrested for stealing bags of U.S. Mail from trains in New Haven, and for that he went to jail.

To Charles Jackson the Tuckerman case showed only that Morton was unchanged since his days in Rochester and Cincinnati and that the congressional petition was less a sincere attempt to gain the award than a new racket by which to induce gullible men to give him money.

Massachusetts General Hospital and its individual physicians were thoroughly weary of Morton by the mid–1850s; they had been tired of his greedy bleat almost since the start, yet they had supported him dutifully with public and private bailouts. In 1858 a group of physicians in New York, led by Valentine Mott, took up the cause from the exhausted Bostonians. They enlisted the greatest fund-raising effort yet on Morton's behalf, hoping to raise $25,000 in a nationwide subscription. To initiate the campaign Mott decided that there should be a book to explain both why Morton was a great man and why, due to cruel fate, he had ended up a poor one. The New Yorkers recommended a twenty-eight-year-old doctor with a background in writing to go to Wellesley and write the book in the summer of 1858. The doctor, Nathan Rice, did as he was told, and completed the book right on time, September 1. His working relationship with Morton was such that he devoted two months to writing the manuscript and then four months to fighting over it—and his payment. Morton felt no particular obligation to his author in either matter. Rice demanded three things: the right to make corrections to Morton's rather impulsive additions, and the money, of course. Third, he insisted that his name not appear in any connection with the book, which Morton had titled *Trials of a Public Benefactor.* Rice loathed the title.

Nathan Rice had studied the controversy at close range during his summer in residence at Etherton Cottage. He had been granted full access to Morton's original documents and close contact with the hero himself. That is why so many people paid close attention to an article he published on the controversy in February 1859, simultaneous with the launch of the book. In the article Rice declared Horace Wells to be the rightful discoverer of surgical anesthesia. Morton, he wrote with a streak of self-recrimination, "evidently considers the pen as mightier than the sword and makes up by multiplicity of documents for weakness of proof."

Morton owed a couple of hundred dollars to young Nathan Rice; he owed tens of thousands of dollars to the Eastern Railroad, and he owed amounts that fell somewhere in between to many other people, including his neighbors in Wellesley.

In Wellesley there was an old legend that a man returning home after delivering a load of wood stopped in the town square, gathered up three branches that were left in his wagon and planted them in the ground, where they grew into a fine, tall line of trees: two buttonwoods and an elm in the middle. One Sunday soon after the publication of *Trials of a Public Benefactor,* William Morton was walking along Washington Street, through the town square, when he noticed that he was also swinging from a low branch on the taller of the buttonwood trees there: hanging in effigy to the delight of his neighbors gathered underneath. "The effigy," according to a writer for a local magazine called *Our Town,* "which bore the name of 'Morton' on the hat band, and was said to be a good representation of the man, was hung on the tree on Saturday night and remained for a week, when it was taken down and burned." Morton walked all around the effigy, mixing among those others it had drawn to the square, and then he nonchalantly lit a cigar. The writer for *Our Town* compared Morton to the man "who, speaking of his wife's striking him, said it gratified her and didn't hurt him." The only people who mattered in Morton's mind were in Washington, and the only news that mattered was news that went to Washington. The buttonwood effigy was of no consequence.

In 1861 the news in Washington, of course, was the breakup of the federal Union: the Great Rebellion, the Civil War. Morton watched with curious interest as the Southern states left the Republic. Of course he would have loved to see Connecticut join the Confederacy, to rid Congress of the Wells faction, but the loss of Georgia and all those sympathetic to Crawford Long was certainly gratifying. Even aside from such partisan hotspots, the change in the makeup of Congress represented a strategic opportunity in Morton's campaign there. In 1861 the old team centered at Mass General resurrected the petition for one hundred thousand dollars and presented it to Congress with new hope. However, the effort failed, because some representatives noted the fact that officially Morton's patent, Number 4848, had expired in 1860. Mainly, though, there was a more obvious reason why Washington failed Morton yet again: The war got in the way.

Bitterly Morton turned on the offensive, using the courts for the first time. The lawsuit he filed against the New York Eye Infirmary in January 1862 was intended to show that the patent was indeed still worth something—that every institution that had ever used ether without his written permission now owed him money. The judge in the case cut the testimony short, however, ruling that the patent had been void since the day it was issued, since it sought to protect a new use for an old compound, not a new invention. With that the case was closed.

During the Civil War anesthetics were used regularly at hospitals on both sides of the conflict. "How much have the horrors of the battle-field and the hospital been diminished by the use of Ether and chloroform!" wrote a surgeon in January 1864, looking back over his years of service. "I remember a poor fellow, a rebel prisoner," he went on, "who at the battle of

Antietam was struck by a ball which passed through his thigh." The limb had to be amputated. "When I told him of the sad necessity," the surgeon recalled,

> he burst into tears, crying, "Oh, doctor, for God's sake save my life, for I am not fit to die!" . . . We gave him ether, and I proceeded with the operation as rapidly as possible. For some time it seemed as if the sleep of anæsthesia would be his last, and I confess that I felt no inconsiderable relief when finally he put out his hand, and smiled, and called me by name again.

"The operation was done in the afternoon, and before midnight he was a corpse," the surgeon added rather distantly.

Another Union surgeon recalled a more successful operation on a colonel for the Union side who was shot in the forearm at Port Hudson, Louisiana, in 1863. The surgeon had organized a hospital as near to the fighting as he dared: a dim, leaky tent that he had tried to hide in the woods. "There were seldom cries or shrieks," he reported, for ether was in use, "few sounds save low, abrupt directions, short and pointed but not unkind questions, and repressed groans." The colonel who was shot was William F. Bartlett, a twenty-two-year-old from Massachusetts. He had lost a leg in battle in 1862 but returned to command again, riding on horseback in order to lead his men. For that reason he made the surgeon promise not to amputate the arm without telling him first.

"So he drank of the oblivion," the doctor recalled years later, "and ceased to suffer, but his dream was not of home. 'Doctor,' he muttered (talking in his ether sleep), 'that's my bridle hand, you know. Never can ride at the head of my regiment if you take that off.'" As it turned out, the bullet came away easily, and young Bartlett's arm was saved. He was later wounded two more times, once leading troops at the Battle of the Wilderness and once leading them at the Crater (Petersburg, Virginia), but the indestructible young colonel survived the war.

By 1862 William Morton was forty-three years old. He had been turned away by Congress in four concerted attempts to prevail with his claim for one hundred thousand dollars: 1849, 1852, 1858, and 1861. His efforts to rally support in between were unabating, driving first his dental practice to ruin, then his manufacturing company, then his farming activities. Morton was obsessed. If his character can be assailed because he had so often looked for lazy, lying, or shortcut means of garnering money, then he must be admired for the tenacity and resourcefulness with which he pursued his hundred thousand dollars. And there had been no shortcut, oddly enough. Morton was said to have been earning twenty thousand per year as a dentist and manufacturer in 1846; by 1861, at that rate, he would have earned three hundred thousand. Instead he was in debt, and he was desperate.

Morton should have given up. That would have been a damning thing to say, an ugly and certainly an antithetical thing to say to a man who had been imbued with the commercial possibilities of his nation during the 1840s. By the time it became clear that Morton should have given up—that he was only ruining his life by keeping on—he apparently couldn't stop.

In the midst of Morton's 1863 campaign in the halls of Washington, the surgeon general surprised everyone by recommending that Morton receive two hundred thousand dollars in recognition of the benefits of inhalation anesthetics. That merry thought reinvigorated Morton's efforts and inspired him to enlist new constituencies. First he decided that the way to reach representatives most effectively was not, in actuality, through personal lobbying in suites at the Willard but rather through the electorate that put each man in Washington in the first place. And so Morton printed circulars and organized mass mailings, requesting leading citizens in communities throughout the country to send letters to their representatives—letters insisting that Morton receive his due reward.

Anesthetics were perceived as one of the few blessings in the Civil War. To correct that impression slightly, however, such that the gratitude fell to the man and not the notion, William Morton went to war and wrote about it in one of his circulars.

"I have observed his course for the last three winters at the Capitol," a Harvard Medical School graduate named Horace Thurston wrote from Washington in January 1864.

"In that circular," Thurston noted, "he claims to be employed as the general, if not almost universal, administrator of Chloroform and Ether. He boasts of being omnipresent in the army; particularly at Fredericksburg .... I never heard of Morton's administering Ether or Chloroform while I was there." Thurston was in position to form a judgment, having been in charge of a surgical field hospital at Fredericksburg.

"I know that he is hoodwinking the unsuspecting," Thurston wrote, and went on to call Morton's circular "a dodge to raise the wind." Out of the confusion of battle, it is always hard to sort out reports and know exactly what any one person did or did not do. With the prejudices of the ether controversy pressed upon recollections, a clear image of William Morton's role is especially elusive. Whatever he did on the field of battle, however, his "never sleeping vigilance" (to use Dr. Thurston's words) made certain that it made news.

A report carried on the Associated Press service in 1864 claimed that Morton had administered ether in two thousand battlefield cases. It went on to detail his methods, with the conclusion that "nothing could be more dramatic, and nothing could more perfectly demonstrate the value of anesthetics. Besides, men fight better when they know that torture does not follow a wound, and numberless lives are saved that the shock of the knife would lose to their family and friends." The source of the information in the article was not hard to trace.

A few weeks before Morton had made many of the same points in writing to his wife, Elizabeth, from Fredericksburg, Virginia, in the neighborhood of fierce fighting in that Spring of 1864. On the same day, May 13, Morton wrote to an associate in Massachusetts: "Here is a copy of a letter I have sent Mrs. Morton," he began:

> It will be of considerable importance to me if you can get it before the public or such portions of it as you approve. You can use my name to it in any form you think best or give it as an Extract. I hope you can get the fact that I am here into the associated press as all the members of Congress and the country would see it.

Morton soon expanded his memoir of his war experience behind the lines at the Battles of the Wilderness and Spotsylvania, and tried unsuccessfully to have it published. In 1904 it was finally printed in booklet form, as a historical curiosity. Probably because there were still those who doubted that Morton had done anything in the Civil War other than mooch at officers' mess tents, the booklet includes a personal recollection from a respected physician who associated with him in the fighting at the Wilderness (a Virginia battlefield). Morton's article, which contains a good-natured description of the battle, as he saw it from the medical tents, concludes with the statement: "For myself, I am repaid for the anxiety and often wretchedness which I have experienced since I first discovered and introduced the anesthetic qualities of sulphuric ether, by the consciousness that I have thus been the instrument of averting pain from thousands and thousands of maimed and lacerated heroes. . . ."

Whether or not he really felt repaid by any such thing, he continued his efforts to secure his reward from the government. The time seemed ripe and the petition, set at two hundred thousand dollars, was directed at the General Appropriations Bill in June of 1864, precisely when reports of Morton's service at the Battle of the Wilderness were timed to detonate in the press. Everything was ready; fifteen years of lobbying had built a veritable architecture of support in Congress and out of it. Thousands of pages of testimony rested behind the single-sheet descriptions of the petition, which hung on the walls of the room set aside for the House Ways and Means Committee. The members duly voted to include it in the appropriations bill. Morton haunted the halls of Congress as the debate in the full House approached, collaring representatives to point out the rows and rows of hospital barracks lined up on the Mall just outside the Capitol; those barracks were filled with men whose lot had been eased by ether. And then the day arrived.

Nothing was as certain as it seemed. Within five minutes the Morton claim had been stricken from the bill, along with a number of other extraneous measures.

William Morton still did not give up.

The claim was sent once more to committee the following autumn, and Morton took to the lecture circuit throughout the country in order to rally support for his plight. John Kasson, a representative from Iowa, wrote to Ralph Waldo Emerson in September. "From the excitement caused by the revival of the subject," he wrote, "we have new evidence of the necessity of such an agent to be administered to rival claimants and to legislators!"

While the petition dragged around Congress, Morton dragged around the nation, speaking to any group that would listen. As time went on, his speeches were less rallying cries than cries for help.

Close to twenty years of fighting for supremacy in the ether controversy had turned William Morton into something of a joke, a man looking much more like a spent mountebank than a man of science. He lived by pawning the medals he'd received from foreign governments, and then redeemed the medals by making lectures before medical societies and other groups. The lectures were neither popular nor informative, but most of the nation's doctors felt that they owed William Morton something. The discovery of anesthetics had, after all, relieved the entire profession, no less than any individual patient, of endless pain. To connect himself to the people to whom he spoke, Morton presented himself as a martyr, a man whose life had been destroyed, even though there was still breath in his body.

"When we consider this discoverer," wrote a schoolteacher who attended a lecture by Morton in 1865, "health impaired,

William T. G. Morton, 1854.
(Van Pelt Library, University of Pennsylvania)

business destroyed, property gone, embarrassed by annoying debts, 'inaction' seems 'crime.'" That schoolteacher, writing in *Harper's*, had had only the vaguest knowledge of Morton beforehand.

Describing William Morton at forty-seven, the teacher wrote: "On the stage by the side of the Principal, was seated a man with a thoughtful, perhaps sad, face and an intelligent blue eye."

Morton impressed the *Harper's* writer. "It was not his polished style that did the work, for Dr. Morton was not at all times fluent." Instead it was the very *fact* of Morton: "I was face to face with a man to whom the suffering every where are indebted, and whose name has been for a quarter of a century ringing in two hemispheres."

Unfortunately it was ringing in controversy. "The ether controversy"—the words themselves became a catchphrase of the era, the ugly residue of a shining triumph. The debate was still raw more than twenty years after the first introduction of anesthetic ether in surgery.

A physician in Chicago, who saw Morton later that year, formed a different impression than the schoolteacher did. He thought Morton an embarrassment, describing how "not a word escaped his lips by which the most attentive listener would have supposed that either HORACE WELLS or C.T. JACKSON ever lived." At the end of the speech in Chicago, Morton invited the audience to look through the many publications he had laid out on a table for their perusal and then to sign an appeal directed toward the most affluent citizens of their city, demanding remuneration for the ether discoverer.

The difference between the two views of Morton's lecture is that the teacher was too young to remember the ether controversy personally. The physician, on the other hand, remembered enough to believe that Morton had actually done more than anyone else to *hinder* the proliferation of anesthetics, that his charlatanism and avarice had actually discouraged many practitioners from embracing etherization at first. Morton's future attempts at securing his reward lay with the new generation, with people such as the schoolteacher, who had no opinion and would listen to a wonderful story—"with an interest which few novels have awakened in my mind," as the teacher expressed it.

In the years following the Civil War, Morton drummed on. The effort was somewhat withered for lack of fresh money and energetic assistance, both of which had once been much easier to procure. The claim was nominally still at issue in Congress, and so Morton toured the nation, constantly looking for help and a new, unjaundiced audience. The field of the ether controversy, once so hectic and crowded with opinion, seemed to be Morton's to fill with his own desperate speeches and his own halting voice.

In June 1868, a new article appeared on the discovery of etherization in the siren voice of the *Atlantic Monthly* magazine. The *Atlantic* was not only the most respected magazine in the country, it was the most respected one in Boston, Morton's own home city and the place where the discovery had occurred.

The article was concise, four pages in length, and the second sentence stated straight out: "The discovery was made by Charles T. Jackson, M.D. of Boston. . . ."

What would the schoolteacher think?

On the third page the article finally—and for the only time—mentioned William Morton, calling him "a dentist of little medical and almost no scientific knowledge." That point may be arguable, but the article's fuller point was simply untrue:

> Dr. John C. Warren, who performed the first surgical operation on a patient under the influence of Ether vapor, on learning that Dr. Morton, a dentist of little medical and almost no scientific knowledge, in administering the Ether vapor in its early applications as an anæsthetic at the [Massachusetts General] Hospital, had acted under Dr. Jackson's directions, expressed his satisfaction that the discovery of etherization had had "a scientific origin."

John C. Warren was never concerned with the origin of the discovery. He may have called Morton "quackish" behind his back, but he accepted Morton's central role in the discovery and never chose the path taken by many others: that whatever the truth, the discovery would "look" better with a respectable Harvard man such as C. T. Jackson beside it. No matter the motivation, though, whether partisanship or sheer cynicism, those lines in the *Atlantic Monthly* are worth rereading, for they are the words that killed William Morton.

19

POISON IN THE BEAMS

It was some time since anything of the sort had appeared," wrote Elizabeth Morton of the *Atlantic Monthly* piece, so authoritative in style, so dismissive of her husband. "This article," she recalled, "agitated him to an extent I had never seen before."

The article had not been out long when William Morton left Etherton Cottage in Wellesley and traveled alone to New York. He intended to publish a response to the *Atlantic Monthly* article in a magazine such as *Harper's.* His old friend Henry J. Bigelow helped by marking up a copy of the *Atlantic Monthly* article, noting the errors it contained. Despite being debt ridden, Morton took rooms at one of the most ostentatious of the city's hotels, the St. Nicholas, and tried to call on those who might help. It was a terrible time to do business in New York, though.

The weather was hot, even for July, and Morton couldn't find the people he needed to see. After just a few days he was exhausted and sick, suffering an attack of rheumatism in his leg, and something he called "brain fever." On July 11 he sent a telegram to his wife, asking her to come take care of him. By the time she arrived, the next day, he was under the care of a number of different doctors.

Morton grew better, but the weather grew worse, taking a steep turn on July 13 to temperatures of more than one hundred degrees.

Throughout New York City people were dropping from sunstroke, as rays beating down from a cloudless sky carried "poison" in their beams, in the opinion of one reporter. He may have been joking, but the *New York Times* cited sunstroke statistics for the previous fifteen years and found something quite akin to poison in the beams:

> Some very warm months have given but a small mortality from this cause [sunstroke], while in others the number of deaths appears to be excessive, even when the temperature does not much exceed the average. What these causes are, which combine with heat to produce fatal congestion, are not precisely known; but heat by itself does not seem adequate to produce these results.

The air was windless for days. Some imbalance in the composition of the atmosphere undoubtedly exacerbated the heat. One woman suffered sunstroke while aboard a ferry and fell into the Hudson River, to be fished out and saved. Some people died where they worked, including a saloonkeeper who died while tending bar and a postal worker sorting mail. Many people died in their sleep, as the heat barely abated at night.

On Wednesday, July 15, Morton's doctors visited him at the St. Nicholas and found that he was growing more disturbed and irrational. They advised him to remain quiet at the hotel, stifling as it was, because there was no room in the city hospitals. That evening after dinner, however, Morton insisted on leaving. He had the idea that he wanted to move to a hotel in Washington Heights, at the northern tip of Manhattan Island. Elizabeth tried to dissuade him, but he would not remain where he was, and so she made arrangements for the rental of a small carriage. While she did that, Morton prepared for the ride, carefully pinning onto his coat the three decorations that he had received from foreign governments: the Order of Saint Vladimir from Russia, the Cross of the Order of Vasa from Sweden and Norway, and a gold medal representing the

Montyon Prize from the French Academy of Sciences. After dinner, at about eight o'clock, he and Elizabeth started out: a small, pretty woman and a tall man jangling his medals.

The pleasantest route north, and the quickest, was through Central Park. After the Mortons entered the park, though, William became drowsy and had a hard time driving. Elizabeth offered to take the reins, but he held on to them. He refused to talk about turning back. The trip through the park took over an hour, and it was 9:20 when they finally arrived at the Harlem Gate, leading out to 110th Street.

"Just as we were leaving the Park," Mrs. Morton later recalled, "without a word, he sprang from the carriage, and for a few moments stood on the ground, apparently in great distress." Not knowing what her husband would do next, and intimidated by a crowd of strangers who gathered to look at the strange picture he presented, Mrs. Morton went to him and took charge of his watch, his money purse, and the three medals he wore. Suddenly Morton became agonized. He bolted from his wife and ran wildly to a pond on the nearby ground of the park, where he plunged his head and upper body into the water.

Mrs. Morton screamed, and two men came from the streets to help: One was Joseph Swan, a doctor, and the other was a policeman named Thompson. They pulled Morton out of the water and immediately recognized that he was suffering from a catastrophe much more complex than sunstroke. Dr. Swan ordered that he be transported to a hospital immediately. Because of Morton's condition a large "double-carriage" had to be summoned, and it was an hour before one arrived. "Two policemen lifted him tenderly upon the seat," Mrs. Morton wrote, "I being unable to do anything from the condition I was in: the horror of the situation had stunned me, finding myself alone with a dying husband, surrounded by strangers, in an open park at eleven o'clock at night."

Officer Thompson held Morton in his arms while the carriage made its way to St. Luke's Hospital, but even before it stopped, Dr. Morton was dead.

At the hospital the chief surgeon examined the body routinely and then recognized the face. "Young gentlemen," he said to the students accompanying him on rounds, "you see lying before you a man who has done more for humanity and for the relief of suffering than any man who has ever lived." It was the sort of tribute Dr. Morton would have savored.

Mrs. Morton heard the surgeon out. She took the three medals that had belonged to her husband and placed them beside his body, while the others watched. "Yes," she said in response to the chief surgeon. "And here is all the recompense he has ever received for it."

The next day the *New York Post* carried a long list of the previous day's victims of the heat wave and listed among them:

"Professor W. T. G. Morton, of Boston, found insensible at One hundred and tenth street and Sixth avenue, and died on the way to St. Luke's Hospital."

It was an easy day for dying in New York City, but the official inquest confirmed Dr. Swan's original diagnosis, that Dr. Morton had not died of the heat, as had so many others on the long list of the stricken, but of "congestion of the brain."

At the age of forty-eight William Morton was just what he had been at twenty-six: not a great man at all, but a man who had given the world a great discovery. He never understood the difference, and on one close night at about eleven o'clock, at a corner on 110th Street, he finally gave up trying.

20

IN DR. JACKSON'S SCRAPBOOK

During the ether controversy Charles Jackson kept a scrapbook, which is now at the Houghton Library of Harvard University. The book measures about fourteen inches tall and nine inches wide, and into it he glued the clippings that told the story of his war for recognition. The scraps have made it a plump book, cinched together with a long shoestring. There is no particular organization to the book; it isn't even strictly chronological, but it casts a view into thoughts that once colored the mind of Dr. Charles T. Jackson.

AFFIDAVIT, headlines one hand-copied note, dated April 2, 1849:

> I do hereby certify that some time in the winter of 1846–7, Mr. W. T. G. Morton offered me, as a bribe, a gold patent lever watch if I would make a false statement to be used in evidence with regard to some business transactions between him and one Dr. N. C. Keep, and that I indignantly rejected his proposal . . . . [signed] I. E. Hemmenway.

"A portrait of Dr. Charles T. Jackson, the discoverer of the pain subduing quality of ether vapor, by Mr. Moses Wright, an eminent artist of this city, has been placed on exhibition at the store of Mr. Childs, No. 19. Tremont Row," began one undated clipping, which went on to note that the same artist had recently returned from Germany, where he had completed a commissioned portrait of Alexander von Humboldt.

The *National Police Gazette*, a base publication that one might think would lie below the reading level of Charles Jackson, gained quite a lot of space in the scrapbook. One editorial, presumably published in advance of the Civil War, found a subplot in the actions of certain figures associated with Morton's side. "[Nathaniel] Bowditch, [Richard H.] Dana and a few others who have given Morton's spurious claims what little consequence they have attained in this city," asserted the newspaper, published in New York, "are of the abolitionist school of politics and men whose judgments are entitled to all the confidence that appertains to such crack-brained enthusiasts. They have the unenviable distinction of being 'backers' of this Morton and having once very foolishly committed in his behalf, are too obstinate to back out."

There are many more clippings about William Morton, including a squib from the *Boston Evening Transcript* of March 29, 1856: "Personal. Dr. Morton, of anaesthetic celebrity, is lying dangerously ill at Willard's Hotel, Washington."

In the midst of all the pieces of the story that are pasted into the book, the painful shards and the silky ribbons, is a clipping glued to the top of a page, bearing the headline JUST THE SAME.

> One of the old philosophers used to say that life and death were just the same to him. "Why then," said an objector, "do you not kill yourself?"
>
> "Because it's just the same," replied the philosopher.

Charles Jackson thought that was worth saving.

As a young man, Jackson had been slight, almost wan in appearance, and prone to illness from almost anything around him: any encroachment, even a change in temperature or humidity. By the time he was in his sixties, though, Charles Jackson was no longer slight; he was a husky man, bearded and often disheveled, as though at last insulated from some part of the world and its encroachments. When he was out in public, he took care, just as Morton always had, to wear the medals he'd received from foreign governments in recognition of the discovery of etherization.

Through most of the years of controversy, Jackson had a comfortable home life. The income from his inheritance augmented his professional fees, allowing him to provide his family with a happy way of life. As did other families in Boston's upper middle class, the family summered at the shore, with Charles visiting every weekend, at least. His niece, Ellen T. Emerson, described the house the Jacksons rented one year at Cohasset: "a most antique and primitive house," she wrote, "large, comfortable, and entertaining but without carpets and almost without furniture, just what I like, and situated on rocks high and wild, overlooking open sea, perfectly magnificent."

Susan Jackson took a great deal of the responsibility for raising their six children, since Charles was away on geological surveys so much of the time, sometimes for months on end. Occasionally, though, Jackson would take one of his three sons with him in the field.

In mid-April 1861 he vexed his twenty-two-year-old niece Ellen Emerson and his nephew (her cousin) Haven Emerson, who were waiting at the Jackson home in Boston for their cousin Lizzie (young Lidian Jackson). "What should Uncle Charles take it into his head to do, but to read aloud to Haven and me one of those wretched croaking 'treasonable' articles in the *Courier*," Ellen wrote home.

The Civil War having just begun the week before, Charles was reading a piece railing against it. "Haven attempted to listen with some show of respect," Ellen wrote,

> but I couldn't stand it, and remembering that propriety forbade stamping, I tried to keep still, but couldn't do it, so after laughing with rage as a safety valve in which Haven was

> thankful to join, I implored Uncle Charles to stop, but he read perseveringly on. As a last resort, I jumped up and walked about with ears eager for some sign of Lizzy's approach, and when I flew to open the door for the sake of doing something, to my immense relief there she was, and we all three talked as fast as we could, while Uncle Charles read louder and louder till at last we prevailed on him to stop long enough to direct us on our way after which we fled.

Uncle Charles was either a terrific tease—which would not be an unusual thing for any person's uncle to be—or he was beginning to rave. The next month, in any case, he charmed his Ellen by giving her some blue writing paper and purple ink, so that she could write an impressively fancy letter home to her family in Concord.

The Civil War did not affect Charles Jackson's activities as a chemist. In 1865, just as the war was ending, he accepted a job surveying the mining region along the California-Nevada border. He traveled to California by boat, making notes on shark behavior along the way. Taking a side trip from San Francisco in order to investigate rumors of oil fields in Santa Barbara County, he camped near Ojai one night in July, 1865. Having studied meteorites all his life, even collecting a few, he had the genuine thrill of seeing a meteor pass overhead in "a burst of intense light," as he called it.

In 1868 Jackson went on his last adventure, a trip to the Canadian Rockies. When he returned home he learned that there had been trouble in his family. A cousin in Plymouth had gained control of the investments that produced income for the three Jackson heirs. Along with his two sisters, Charles Jackson turned to Waldo Emerson for help. Emerson enlisted advisers and tried to sort through the ravaged inheritance, but the problems compounded. The relative succeeded in seizing most of the property in the old inheritance.

Charles Jackson found himself in reduced circumstances. He turned down the presidency of his cherished Boston Society of

Natural History in 1870 but still attended its meetings. He tried to retain his connections to the city's scientists. At sixty-five he was an elder in a princely generation of Bostonians, intellectuals of vigor who had, among other things, turned Harvard from a local college into a great university, and then founded a dozen other respected schools. They valued nothing so much as a good mind, that generation to which Charles Jackson rightly belonged.

Yet the princes of Boston's own hierarchy left Charles Jackson behind. Within Charles's family, there was a rustle of excitement after Oliver Wendell Holmes, who never publicly took a stand in the ether controversy, introduced Charles Jackson at a party, calling him the "discoverer of Etherization." Lidian Emerson wrote all about it to her husband, who was away on a lecture tour at the time. She was ecstatic over such a small thing: "Charles told me that Dr. O.W. Holmes at a party at the Revere House—brought Mr. Reade—nephew of Charles R. and introduced him to the Dr. as the 'discoverer of Etherization!!' "

Privately it was different, though. Holmes was much less pleasant, and less politic, where Emerson's brother-in-law was concerned. In a letter to a close friend, he wrote of Charles Jackson, "Although it is clear as day that *J* jumped up behind, he is a good natured fellow enough, and one does not care to tug at his little red ribbons and other 'crachets' as the Frenchmen coarsely call those decorations.

"But if the question is publicly agitated once more, my testimony and opinion will be where they always have been, with Mr. Morton," Holmes added.

Long after the fight should have been over, Jackson was wearing medals that his old friend would coarsely call *crachets*—drops of spittle, in the literal translation.

In 1872 there were two big pieces of news in the Emerson family, neither of which directly concerned Charles, who worked quietly in his own lab that year, trying to fend off financial problems. The big news was a fire at the Emerson house. It damaged

Ralph Waldo and Lidian Jackson Emerson (center) *with children and other family members in front of their home in Concord, Massachusetts, c. 1870.* (Concord Free Public Library)

only the upper floors, but the strain of fighting it, shoulder to shoulder with neighbors from Concord, left Emerson diminished. Later in the year he took his daughter, Ellen, on a tour through Europe and Egypt, while Lidian stayed home in Concord. The vacation represented no breech within the couple—in her whole life Lidian Emerson had never consented to any trip beyond the environs of Boston. An overseas voyage was out of the question, but on May 28, 1873, Lidian was waiting at home when Waldo and Ellen arrived back in Concord.

Most of the rest of the town was down at the train station, greeting the three o'clock express from Boston. Actually, it was more than just *most* of the town. It was "the whole population," in Ellen's estimate, "horse, foot, wagons, babywagons and schools all in array at the depot." She and her father walked past a wagon loaded with his grandchildren, and then climbed into a carriage to join the procession, which escorted them about one mile to their house, where Lidian met them at the door. That was how Ralph Waldo Emerson came home, when he came home from his last long journey.

In reciprocation the Emersons invited the whole population of Concord to an open house, scheduled for Saturday, June 22, three weeks later. Two days before the reception, Uncle Charles and Aunt Susan arrived in Concord with their daughter, also named Lidian, to stay at the house. "Uncle C., Aunt S. and Lidian are going to stay till Saturday and keep Uncle C.'s birthday," Ellen wrote to her sister. The next day, June 21, Charles Thomas Jackson turned sixty-seven. It was a personal holiday that he celebrated as he did all the national ones: with his sisters, all of their families together at the Emersons'. Later that day he and Susan took the train home to Boston, while Susan was to return for the big party the next day.

Charles's daughter, known in the family as young Lidian, stayed in Concord to help her cousin Ellen prepare the house for the June 22 reception. Everything had to be "perfect, top to bottom," in Ellen's estimation. She was so carried away by excitement that she finally stalled out completely on Sunday morning, staring at the dining table and trying to arrange the serving pieces in exact symmetry. She couldn't get it right, no matter what she tried. Finally she called her clever friend Louisa May Alcott, who had been in the house all morning helping with the preparations. "She squinted learnedly at the problem for a minute," Ellen reported, "suggested the simplest possible change, and that being made, brought all right. 'That is what I call applying MIND to MATTER,' cried she triumphantly."

Young Lidian, for her part, was put to work mixing large batches of lemonade. In that she was far more capable than Louisa May Alcott, who didn't bother to use water, only lemon juice and sugar, and so was the only one who could drink her own concoction. The reception was scheduled to begin at four that afternoon.

At noon a telegram arrived for Lidian the elder. It was from Susan Jackson, still in Boston. Charles had fallen down in his study and was unconscious.

Lidian was distraught at the thought of her little brother—her rather big and worn and sadly old little brother—overcome and verging on death. Others consoled her that Charles had been taken ill many times before, with shocks and breakdowns from which he had always recovered. Perhaps he only needed to rest, she was told.

Four hours later the reception came to life at the Emerson house. "Mother was anxious about him," Ellen wrote, referring to her Uncle Charles, "but we had the reception just the same. My friends were marshalled in Committees, one to show people over the house, one to take them to the garden, and one to invite them into the dining-room, one to wait on them there. Mother & Father and I stood in the parlor all the time. It seemed to us very pleasant."

At six, when the party ended as precisely as it had begun at four, young Lidian Jackson boarded a train for Boston. When she arrived home on Somerset Street, her father was still unconscious. The doctors called it paralytic shock. That evening he awoke, and his wife was overjoyed to watch as he drew himself, by himself, out of bed to walk across the room. Silently he examined his own face to see if it was contorted. It was—it was twisted all out of shape. He couldn't speak at all. But he could walk, and the doctors were optimistic that with time, every aspect of the paralysis would recede.

The next morning Lidian Emerson arrived to see her brother. In the middle of the week Emerson arrived to visit, and Charles began to speak—yet he formed words that weren't intelligible. On Thursday, Susan wrote to Lidian, with stark news: Charles was not getting better, he was getting worse. Lidian visited again on the weekend, and then she knew what Susan did, saw what she saw.

Charles could speak clearly but didn't use English, only words no one had ever heard before. He could neither take care of himself nor allow anyone else to do anything for him, and so he was taken by force to the McLean Asylum to be treated as insane. Located just outside of Boston, the asylum was part of Mass. General. By early August, Susan Jackson was writing to Waldo Emerson, "I have no hope at all of his recovery."

And one week later: "I have no hope for my poor husband . . . ."

Whenever Charles's condition and behavior allowed him to be presented to visitors, the asylum would inform the family. On those occasions Susan and Lidian would usually accompany each other to Somerville, where the McLean Asylum occupied a campus of granite buildings. On every visit Charles was thrilled to see them, overjoyed and overflowing, talking to them almost without stopping from the moment they arrived until they finally left. "But they could understand nothing," Ellen noted.

Lidian only knew that her brother was homesick and unhappy. "And then," Ellen added in describing her mother's visits, "what was it that he seemed so earnest in telling them?"

In late November, Susan Jackson received a letter informing her that the board of trustees of Mass. General had voted unanimously to underwrite Jackson's care at the asylum, "in recognition of the services of Dr. Charles T. Jackson in connection with the discovery and use of Ether at the Mass General Hospital; and furthermore, remembering his many kindnesses during the past years to the patients of the McLean Asylum."

Charles Jackson was a ward of the McLean Asylum, isolated from all—all that had gone before. Perhaps never isolated, though, from what one old friend called a "life embittered by a lack of appreciation for his services to humanity and by ungrateful indifference to his merits."

In 1880 Dr. Charles Jackson died at McLean after surviving for seven years: a man seen in one world and yet seeing another. The Boston newspapers reported his affliction only as "a long illness."

A colleague in the Boston Society of Natural History, Professor T. T. Bouvé, spoke more bluntly to the society about

Charles's death: "Not in sadness, alas, did the friends of his earlier years learn of the final departure of him whom they had respected and loved," he said. "Their mourning had been a continuous one from the moment when the bright intellect that drew them about him in wondering delight, had fallen into an eclipse, never again to manifest on earth its wonted power and brilliancy.

"And to them," he said, "the news of his death came as a relief.

"The troubled spirit had at length passed from its frail and disordered tenement," Bouvé continued, "and the peace which rested upon the form and features of the dead, diffused itself and permeated the hearts of many who had long waited the final consummation." Blunt, just the way Jackson would have wanted him to be, Bouvé mourned Charles Jackson's life, not the seven years of his death.

It was not merely that Horace Wells, William T. G. Morton, and Charles Jackson each came to the day for him to die. It was that each one came to a moment when he could no longer live, when an agony of life came to quiet. Their lives had made a sharp turn long before, from comfort and promise toward that moment. Had they lived according to their earlier plans, there might have been no madness unveiled, no suicide, no eclipse. But their lives did turn, and each at exactly the same moment: at a little after ten-thirty on the morning of October 16, 1846, a day celebrated as Ether Day, when anesthetic gas was first demonstrated in a surgical operation. On that day, the mind of man had an answer to pain.

But pain was not through, and with some sort of vengeance, pain itself also turned sharply that day, to conquer the men who dared to conquer it.

EPILOGUE

SINCE ETHER DAY

The first anesthesiologists learned to harness the very power of sleep. In utilizing nitrous oxide, ether, or chloroform, they discovered a passageway leading to a specific band in the states of mind: an insensible sleep known as narcosis. Their successors managed to explore that shadowy band, charting each individual patient's course through it, so that, over the past 155 years, anesthesiology can be said to have been fully realized as an instrument of medicine— to have been mastered.

Without ever having been understood.

Anesthesia has two basic parameters, and they just so happen to coincide exactly with the two most common fears regarding surgery: an application too strong, leading to death, and one too weak, leading to premature consciousness—the patient's waking up in the middle of an operation. In the ongoing effort to avoid either case, the nineteenth century made advances on two fronts. The first improved the delivery system for anesthetic gases. The handkerchief soaked with chloroform was replaced by cone inhalers for ether after 1850, and pressurized containers for nitrous oxide after 1868. By these means, especially the latter, the level of narcosis could be stably maintained through a series of weaker doses, "interval narcosis," rather than one strong snort at the beginning. The second improvement was made in the close observation of the anesthetized patient's vital signs. Long considered an advantageous luxury, this attentiveness became widely practical late in the nineteenth century, when administration of anesthesics shifted away from the surgeon or his disinterested assistant, and became an accepted specialty for both doctors and nurses.

Taut control over delivery systems and expertise in monitoring brought the most common anesthetics a long way toward safe application in the early years of the twentieth century. Even so, none of them was entirely acceptable: Ether was highly explosive; chloroform had certain toxic properties. The complaint against nitrous oxide—heard ever since 1846—was that it wasn't quite strong enough. In 1929 a concerted effort resulted in the advancement of cyclopropane as the first "modern" anesthetic. Others followed; Halothane, introduced in 1956, became the most prominent of all. Inducement of anesthesia today rests on a sensitive recipe designed for each patient, but many operations still rely on the first anesthetic, Wells's nitrous oxide, augmented by one or more other, stronger substances, such as halothane.

For all of its long history, anesthesiology has seen great refinements in just the past generation. The death rate from anesthesia dropped from 1 in 4,500 in 1970 to 1 in 400,000 twenty-five years later. And even that figure is only an estimate, because so few cases occur anymore. Fear of anesthetic death, the lower parameter in the band of narcosis, has been rendered unjustifiable—from the statistical point of view, at any rate. The other great fear, that of waking up prematurely, is still pertinent, however. About 1 percent of patients claim to recall becoming conscious during surgery.

Modern textbooks on anesthesiology run to the thousands of pages, yet for all of the minute understanding that has accumulated, no one really knows why anesthetics work.

The first serious theory was postulated in 1847, just months after ether was introduced for use in surgery. A pair of German scientists named Baron Ernst von Bibra and Emil Harless noticed that anesthetic gases were strongly drawn to fat molecules (lipoids) in the human body. They drew unlikely conclusions, but their basic research led to recognition that, of all the systems in the body, nerve tissue has the ratio of lipoids-to-water that is most attractive to anesthetic substances.

In 1909 R. S. Lillie, an American, suggested that since the membrane of a nerve cell is made up of lipoids (along with protein molecules), anesthetic molecules attaching themselves to those lipoids could block the membrane and interfere with cell flow and function. An earlier scientist, thinking along these same lines of anesthesia by clogging, had made note of the fact that the interior portion of cells that had been affected by anesthetics—by drugs as well as natural phenomena such as heat, cold, or electric shock—becomes slightly "gelatinized" and slow to send or respond to impulses. By 1912 attention directed at the cell membrane turned from the lipoids to the proteins. According to the theory, anesthetics, by attaching themselves to the amino acids that comprise the proteins, temporarily denied them their usual gulps of oxygen, and so their energy.

All the theories concluded that anesthetics interfered with *something* in the cell: lipoids, the interior matter, or proteins. A theory developed by Nobel laureate Linus Pauling in 1961 concluded that anesthetic molecules insinuate themselves into water molecules—the brain being 78 percent water—forming microscopic crystals that trap electrically charged particles and ions, and "thus cause the level of electric activity of the brain to be restricted." The primary argument against Pauling's theory was that such phenomena may occur in the test tube, but not in the body, where the temperature of 98.6 keeps any such crystals from forming.

Dr. Vincent J. Collins, author of the massive text *Principles of Anesthesiology,* reviewed these and other conclusions regarding the effect of anesthetics on body functions, and he drew his own firm conclusion: "None of the theories answers all of the various factors of the problem."

Anesthesia—so elusive for such a long time—still has its secrets.

ENDNOTES

Abbreviations

Libraries

Boston Public Library Rare Books and Manuscripts	BPL
Harvard Medical Library in the Francis A. Countway Library of Medicine	Harvard
Medical Library in the Francis A. Countway Library of Harvard Medical School	Countway
Original materials used by permission of the Ralph Waldo Emerson Memorial Association and by permission of the Houghton Library, Harvard University	Houghton
Massachusetts General Hospital, Boston	MGH
Massachusetts Historical Society, Boston	MHS

Publication

Boston Medical and Surgical Journal	*BMSJ*

Prologue: The Laughing-Gas Joke

Hospital description: E. H. Poole and F. J. McGowan, *Surgery at the New York Hospital One Hundred Years Ago* (New York: Paul Hoeber, 1929), pp. 1–2.

Review: N. P. Willis, "Diary of Town Trifles," *New Mirror*, Apr. 6, 1844, 3–1, pp. 8–9.

Remark on science: Untitled article, *New York Tribune*, Mar. 16, 1844, p. 4, col. 4.

CHAPTER 1: ETHER DAY

Heywood letter: Reginald Fitz, "The Value of Imponderables," *New England Journal of Medicine* 236, no. 16 (Apr. 1947): 557.

Josiah Holbrook: Paul W. Stoddard, "The American Lyceum," Ph.D. diss., Yale University, 1947, pp. 63–67.

Joseph Wightman as assistant: Ibid., p. 66–67

Gould background: George E. Gifford Jr., "The Forgotten Man of the Ether Controversy," *Harvard Medical Alumni Bulletin* 40, no. 2 (Christmas 1965): 16.

Gould quote: *Statements, supported by evidence, of Wm. T. G. Morton, M.D., on his claim to the discovery of the anaesthetic properties of ether*, U.S. Senate Select Committee, 32nd Congress, 2d sess., Jan. 21, 1853, pp. 267–68.

Mrs. Morton quote: Elizabeth Whitman Morton, "The Discovery of Anesthesia," *McClure's*, Sept. 1896, p. 312.

Morton quote on Chamberlain: W. T. G. Morton, *A Memoir to the Academy of Sciences at Paris on a New Use of Sulphuric Ether* (New York: Henry Schuman, 1946), p. 18.

Warren on painful surgery: John C. Warren, *Etherization, with Surgical Remarks* (Boston: Ticknor & Fields, 1848), p.1.

Slade on Warren: Daniel Denison Slade, M.D., "The First Capital Operation Under the Influence of Ether," *Scribner's*, July 1892, p. 521.

Thursday Evening Club: *Statements*, testimony of John C. Warren, p. 307.

Lyell on mastodon: Sir Charles Lyell, *A Second Visit to the United States of North America*, vol. 2 (New York: Harper & Bros., 1855), p. 270.

Jackson on bones: John C. Warren, M.D., *Descriptions of a Skeleton of the Mastodon Giganticus of North America* (Boston: John Wilson, 1852), p. 188.

Warren's journal entries: John C. Warren journals, MHS, Warren papers, vol. 82.

Mass General building quote: Harold Kirker, *The Architecture of Charles Bulfinch* (Cambridge: Harvard University Press, 1969), p. 217.

Building expanded: Ellen Susan Bulfinch, ed., *Life & Letters of Charles Bulfinch* (Boston: Houghton, Mifflin & Co., 1896), pp. 195–96.

Slade on hospital: Slade, "The First Capital Operation," p. 520.

Mummy: John C. Warren, M.D., "Description of an Egyptian Mummy, presented to the Massachusetts General Hospital, etc." *Boston Journal of Philosophy & the Arts* 1 (1824): 164, cont. vol. 2.

Warren on Abbott operation: Warren, *Etherization,* pp.4–5.

Delay and Warren jest: Gifford, "The Forgotten Man of the Ether Controversy," 17.

CHAPTER 2: A BLANK WHIRLWIND OF EMOTION

Richerand operation: "Extraordinary Surgical Operation," *American Journal of Science and the Arts* 3 (1821): 373.

Valentine Mott operation: Valentine Mott, "Reflections on Securing in a Ligature the Arteria Innominata," *Medical and Surgical Register* 1, no.9 (1818), reprinted in A. Scott Earle, M.D., *Surgery in America* (Philadelphia: W.B. Saunders, 1965), p. 96.

Struthers quote: Struthers to J. Collins Warren, *The Semi-Centennial of Anesthesia* (MGH, 1896), pp. 92–93.

Pott on operations: *Chirurgical Works of Percival Pott, FRS* (Philadelphia: n.p., 1819).

Farmhouse amputation: J. Collins Warren, "The Influence of Anesthesia on the Surgery of the 19th Century," an address before the American Surgical Association, Boston: n.p., 1906, p.7.

Boott quote: "Painless Operations in Surgery," *North British Review* 7, no. 8 (May 1847): 185–86.

Wardrop bleeding method: James Wardrop, Esq., "Some Observations on a mode of performing operations on Irritable patients," *Eclectic Repertory & Analytical Review* 11, no.1 (Jan. 1821): 24–25.

Velpeau advice: M. le professeur Velpeau, *Leçons orales de clinique chiruricale faites à l'hôpital de la Charité,* 1840, quoted in Roselyne Rey, *The History of Pain* (Cambridge, Mass.: Harvard University Press, 1993), p. 59.

Eyeball removal: John M. Butter, M.D., "Case of Gun-shot Injuries of the Eye," *Baltimore Medical and Surgical Journal* 2, (1834): 480–81.

Wardrop box for children: Wardrop, "Some Observations," 24–25.

Irish bachelor: "Ether and Chloroform," *Dublin Review* (Sept. 1850): 235.

Warren patient's suicide: Warren, *Etherization*, p. 36.

Buffon death: John Ashhurst Jr., M.D., "Surgery Before the Days of Anæsthesia," *The Semi-Centennial of Anesthesia* (MGH, 1896), p. 30.

Syng on lithotomy: "Of Stone in the Bladder," John Dorsey Syng, reprinted in *Surgery in America*, edited by A. Scott Earle (Philadelphia: W.B. Saunders, 1965), pp. 51–58.

Marshall operation: Marshall to Story, Charles C. Smith, ed., "Letters of Chief Justice Marshall," *Massachusetts Historical Society*, 2nd series, vol. 14, pp. 346–47.

Letter to Simpson: Ashhurst, "Surgery Before the Days of Anæsthesia," 31–32.

Liston quote: Warren, "The Influence of Anesthesia," p. 10.

Chapter 3: The Hilarity Before Ether Day

Dr. Coult: Jack Rohan, *Yankee Arms Maker* (New York: Harper & Bros,1935), pp. 26–50.

Beddoes quote: Joseph Cottle, *Reminiscences of S. T. Coleridge and R. Southey* (New York: Wiley & Putnam, 1848), p. 197.

Pneumatic Institute: "Dr. Beddoes' Medical Pneumatic Institution," *Medical Repository* 3 (1800), p. 423

Davy tries nitrous oxide: "Extract of a Letter from Mr. H. Davy, dated April 17, 1799," *A Journal of Natural Philosophy, Chemistry and the Arts* 3 (May 1799): pp. 93, 446–52.

Cottle on Davy and nitrous oxide: Cottle, *Reminiscences of S. T. Coleridge*, pp. 199–201.

Southey quote: R. D. A. Smith, *Under the Influence* (Park Ridge, Ill.: Wood Library-Museum of Anesthesiology, 1982), p. 54.

Harp quote: William P. Barton, *A Dissertation on the Chymical Properties and Exhilarating Effects of Nitrous Oxide Gas* (Philadelphia: n.p., 1808), p.42.

Davy article quote: Humphry Davy, "Researches on Nitrous-Oxide Gas," *Proceedings of the Royal Society*, 1799, p. 566.

Barton quote: Barton, *A Dissertation*, preface.

Poem: Thomas Green Fessenden, *Terrible Tractoration* (Boston: Tuttle, Weeks & Dennett, 1836), pp. 3–8.

"the most eminent of the benefits" description: "New Medical Discovery," *Chambers's Edinburgh Journal*, Feb. 27, 1847, p.140.

Anne Warren letter: Warren to mother, March 20 [1825], Vaughan papers, Warren correspondence 1825, MHS; also, John McAleer, *Ralph Waldo Emerson* (Boston: Little, Brown, 1984), p. 66.

Fairfield Medical College: *Horace Wells Centenary*, edited by William J. Gies (Chicago: American Dental Association, 1948), p. 47.

Silliman cases: Benjamin Silliman, "Two Singular Cases of the effects of nitrous oxide, or exhilarating gas," *American Journal of Science and Arts* 5 (1822): 195.

Jackson on school ether: Charles T. Jackson, M.D., *Manual of Etherization* (Boston: Mansfield, 1861), p. 16.

Cuthbert letter: G. Cuthbert to friend, Feb. 16, 1760, Harvard, Hms misc. Cuthbert.

Mitchell book: Thomas D. Mitchell, *Elements of Chemical Philosophy* (Cinncinati: Corey & Fairbank, 1832), p. 172.

Chapman book: Prof. N. Chapman, *Elements of Therapeutics & Materia Medica* (Philadelphia: Carey & Lea, 1825), vol. 2, p. 274.

Silliman on ether: Benjamin Silliman, "Effects of Inhaling the Vapour of Sulphuric Ether," *Journal of Science and the Arts* 4 (1818): 158.

Gardner Colton recollections: Gardner Q. Colton, *Boyhood Recollections: A Story with a Moral* (New York: E.M. Day, 1891), p. 2; *Boyhood and Manhood* (New York: A.G. Sherwood, n.d.), p. 7; *A Few Selected Letters* (New York: A.G. Sherwood, 1888), p. 5.

Colton advertisement: *New York Tribune*, Mar. 18, 1844, p. 3, col. 5.

Hone diary: Philip Hone, *The Diary of Philip Hone*, edited by Allan Nevins (New York: Dodd, Mead, 1927), p. 515.

Origin of Life lecture: Patricia Click, *The Spirit of the Times: Amusements in Nineteenth-Century Baltimore, Norfolk and Richmond* (Charlottesville: University Press of Virginia, 1989), p. 28.

Colton's Exploding Water: advertisement for Second Grand Exhibition, *New York Tribune*, Mar. 22, 1844, p. 3, col. 4.

Chapter 4: The Entertainment Is Scientific

Wells letter from Hartford: H. Wells to Susan Shaw, Nov. 21, 1835, in W. Harry Archer, D.D.S., ed., "Life and Letters of Horace Wells: Discoverer of Anesthesia," *Journal of the American College of Dentists* 11, no. 2 (June 1944): 89.

Wells letter to parents: H. Wells to parents, Nov. 25, 1836, ibid.

Letter about Charley: H. Wells to mother, Jun. 26, 1842, ibid., 100.

Friend's opinion of Wells: *Discovery of Anesthesia by Dr. Horace Wells: Memorial Services at the 50th Anniversary, Dec. 11, 1894* (Philadelphia: Patterson & White Co., 1900), p. 27.

Letter from Boston: H. Wells to wife [Elizabeth], Oct. 28, 1843, in Archer, "Life and Letters of Horace Wells," pp. 101–2.

Dissolution of partnership: H. Wells to W. T. G. Morton, Nov. 22, 1843, Harvard, Hms c59.8.

Erving recollections: Henry Wood Erving, "The Discoverer of Anæsthesia: Dr. Horace Wells of Hartford," *Yale Journal of Biology & Medicine* 5, no.5 (May 1933): 425.

Clarke recollection: David Clarke deposition excerpted in *Discovery of Anesthesia by Dr. Horace Wells*, p. 103.

Wells's query: address by Gardner Colton, Philadelphia, Dec. 11, 1894, ibid., p. 41.

Riggs on death or success: John M. Riggs to Drs. Harbrouk & Howland, Sept. 17, 1872, reprinted in Max E. Soifer, D.D.S.,

"Historical Notes on Horace Wells," *Bulletin of the History of Medicine* 9 (1941): 108–9.

Riggs description of trial: *Discovery of Anesthesia by Dr. Horace Wells,* p. 15.

Wells quote on discovery: Gardner Q. Colton, "The Invention of Anesthesia," letter to the editor, *New York Times,* Feb. 5, 1862.

Wells being tired: Norman W. Goodrich deposition excerpted in *Discovery of Anesthesia by Dr. Horace Wells,* p. 116.

Cooley's interest: Sam Cooley to W. T. G. Morton, Oct. 10, 1852, Harvard, HmS c59.8.

Wells in Boston: Horace Wells, letter to the editor, *Hartford Courant,* Dec. 9, 1846, p. 2.

Taft quotes: C. A. Taft, M.D., deposition excerpted in *Discovery of Anesthesia by Dr. Horace Wells,* p. 104.

Wells retreat: Morton, *A Memoir,* p. 7.

Chapter 5: Unwilling Collaboration

Wells on Morton and Jackson: Horace Wells, "The Discovery of Ethereal Inhalation," *BMSJ* 36 (May 1847): 298.

Marcy recollection: Deposition of E. E. Marcy, Dec. 1, 1849, excerpted in *Dr. Wells, The Discoverer of Anæsthesia* (New York: J.A. Gray, 1860), p. 9.

Morton's estate: Barbara Corely Teller, "William Thomas Green Morton: Scientific Farmer," *Wellesley Townsman,* Dec. 19, 1981; Jan. 12, 1982; Wellesley (Mass.) Library clipping file: William Morton.

Jackson on Morton's business: C. T. Jackson to G. W. Childs, Jan. 17, 1859, Library of Congress, Manuscript Collection, Charles Jackson papers, MS 66-1424.

Elizabeth Williams: Deposition excerpted in *Discovery of Anesthesia by Dr. Horace Wells,* p. 123.

Morton ignoring Wells: H. Wells to W.T.G. Morton, Jun. 8, 1846, Countway, HmS c59.8.

Morton on Nig and animals: Morton, *A Memoir,* p. 9.

Jackson on Nig: Jackson, *Manual of Etherization,* p. 119.

Mrs. Morton quote: Elizabeth Whitman Morton,"The Discovery of Anesthesia," *McClure's,* Sept. 1896, pp. 311–12.

Whitman sees Jackson: Francis Whitman affidavit of Mar. 25, 1847, "The Ether Discovery," *Littell's Living Age* 201, Mar. 18, 1848, p. 534.

Eddy and the patent: R. H. Eddy to surgeons of the Mass General Hospital, May 22, 1847, ibid., p. 546.

Chapter 6: Silence in the Dome

Mrs. Morton waiting: Morton, "The Discovery of Anesthesia," p. 312.

Mrs. Morton on husband's return: Ibid.

Chapter 7: The Confidence Man

Jackson on Morton's writing: C. T. Jackson to G. W. Childs, Jan. 17, 1859, Library of Congress, Manuscript Collection, Charles Jackson papers, MS 66-1424.

Gould on Morton's writing: Gifford, "The Forgotten Man of the Ether Controversy," p. 17.

Morton's recollection of youth: Sarah Josepha Hale, "Etherton Cottage and the Discoverer of Etherization," *Godey's Lady's Book* 46 (Mar. 1853), p. 212.

Mrs. Hale on Morton: Ibid., p. 206.

Clerkship: T. Chapman, affidavit, Apr. 9, 1849, Charles T. Jackson Papers, MHS.

Rochester church: G. F. B. Hallock, *A Living Church: The First Hundred Years of the Brick Church in Rochester* (N.p.: Henry Conolly, 1925), pp. 6–12.

Ames recollection: of Loren J. Ames, affidavit, Apr. 11, 1849, Charles T. Jackson Papers, MHS.

Cook recollection: Phineas B. Cook, letter to Joseph L. Lord, Apr. 7, 1849, ibid.

Brick Church meeting: David Dickey, clerk of Brick Church, copy of records from Brick Church of Rochester, N.Y., Apr. 15, 1839, ibid.

Seward School: J. W. Seward, affidavit, Apr. 9, 1849, ibid.

Rochester notice: "Look Out for a Villain!" *Rochester Daily Democrat*, Feb. 11, 1841, p. 2.

Falsified connection to governor: George H. Bates, affidavit, Apr. 23, 1849, Charles T. Jackson Papers, MHS.

Creagh recollection: John Creagh, affadavit, Apr. 28, 1849, ibid..

Circumcision: *National Police Gazette*, February 21, 1852.

Bates recollection: George H. Bates, affidavit, Apr. 23, 1849, MHS, Charles T. Jackson Papers, MHS.

Morton activities in St. Louis: J. W. Southack, affidavits, May 7, 1849, ibid.

Sickles letters: Letter to Messrs Sickles & Co., Aug. 22, 1840; and letter to George H. Hartwell, Aug. 28, 1840, ibid.

d'Lange recollection: M. d'Lange, affidavit May 7, 1849; also, M. d'Lange to F. J. Gray, Mar. 10, 1847, ibid.

St. Louis notice: "Beware a Villain," St. Louis *Daily Evening Gazette*, Sept. 30, 1840.

Baltimore activities: Nathaniel S. Jacobs, affidavit Apr. 3, 1847; George Riggs, constable, affidavit Sept. 28, 1849, Charles T. Jackson papers, affadavit.

Editorial on seals: "Imperfections in the United States Laws," *Boston Evening Transcript*, Mar. 25, 1861, p. 1.

Whitman family history: "Mrs. Morton" (obituary), *Boston Transcript*, Apr. 22, 1904; *New York Evening Telegram*, Apr. 22, 1904.

Elizabeth's recollections: Morton, "The Discovery of Anesthesia," p. 312.

Cook's remonstrance: Cook, Charles T. Jackson papers, MHS.

Chapter 8: Next What?

Classified ads: *Boston Daily Evening Transcript*, Nov. 20, 1846, p. 2.

Gould takes patients: Gifford, "The Forgotten Man of the Ether Controversy," 17.

Morton-Wells correspondence: Archer, "Life and Letters of Horace Wells," 116–17.

First meeting of club: Fitz, "The Value of Imponderables," 557–59.

Wells recollection of Boston: W. Harry Archer, D.D.S., *Chronological History of Horace Wells* (Pittsburgh: A.R. Plantz, 1939), p. 5.

Thayer's recollection of compound: "Extract from Dr. W. H. Thayer's Journal," typescript, MGH, MC3, box 9, file 149.

Warren's letter to Morton: J. C. Warren to W. T. G. Morton, Oct. 28, 1846, typescript, Countway, bMS c70.2.

Bigelow youth: Oliver Wendell Holmes, "Henry Jacob Bigelow," *Proceedings of the American Academy of Arts and Sciences* 26 (1891): 340.

Bigelow essay on numerical method: ibid., 342–43.

Holmes's recollection: ibid., 344.

Hayward attitude: "Some Account of the First Use of Sulphuric Ether by Inhalation in Surgical Practice," George Hayward, M.D., *BMSJ*, 36–12, April 21, 1847, p.231.

Chapter 9: Power Struggle

Burbank recollections: Augustus Burbank letter to sister, Nov. 10, 1846, Countway, bMS misc. B; see also Burbank letters of Nov. 30, 1846, Dec. 23, 1846, Feb. 14, 1847, Countway, ibid.

Morton on tension before operation: Morton, *A Memoir*, p. 21.

Slade description: Slade, "The First Capital Operation," p. 521.

Patent issued: Photocopy of patent, MGH, MC3, box 9, file 151.

Warren's ether prank: John Collins Warren, M.D., "Inhalation of Ethereal Vapor for the Prevention of Pain in Surgical Operations," *BMSJ* 36, no. 375 (Dec. 1846): 378.

Everett's address: E. Everett address, Nov. 6, 1848, MHS, C. T. Jackson papers, p. 491.

J. F. Flagg protest: J. F. Flagg, "The Inhalation of an Ethereal Vapor to Prevent Sensibility to Pain during Surgical Operations," letter, *BMSJ* 35, no. 8 (Dec. 1846), p. 357; also, Richard M. Hodges, A *Narrative of Events Connected with the Introduction of Sulphuric Ether into Surgical Use* (Boston: Little, Brown, 1891), p. 59.

Brand-name meeting: Gifford, "The Forgotten Man of the Ether Controversy," 17.

Holmes's note regarding name: O. W. H. to W. T. G. Morton, reprinted in Albert H. Miller, M.D., "The Origin of the Word 'Anesthesia,'" *BMSJ* 197, no. 26 (Dec. 1927): 1221.

Early anesthesia article: John Yelloly, M.D., "History of a Case of Anesthaesia," *Eclectic Repertory & Analytical Review* 5, no. 19 (1815): 349–50.

Jackson's opinion of Bigelow: Hodges, *A Narrative of Events*, pp. 65–66.

CHAPTER 10: A WORLD WAITING

Jacob Bigelow's letter to London: John F. Fulton, M.D., "The Reception in England of Henry Jacob Bigelow's Original Paper on Surgical Anesthesia," *New England Journal of Medicine* 235, no. 21 (Nov. 1946): 745.

***London Medical Gazette*:** Ibid., 745–46.

Liston operation: Ibid.; Thomas W. Baillie, *From Boston to Dumfries, The First Use of Anesthetic Ether in the Old World* (Dumfries, Scotland: Robert Dinwiddie, 1966), p. 9.

Liston letter to Boott: Richard H. Ellis, M.B.,"Robert Liston's Letter to Dr. Francis Boott: Its Reappearance after 135 Years," *Anesthesiology* 62, nos. 331–35 (1985): 331.

Letters to the *Lancet*: Fulton, "The reception in England," 746.

Fraser use of ether: Baillie, *From Boston to Dumfries*, pp. 13–21.

Edinburgh professors: "Painless Operations in Surgery," *North British Review* 7, no. 8 (May 1847): 176.

Velpeau on etherization: Henry J. Bigelow, M.D., *Ether and Chloroform: A Compendium* (Boston: David Clapp, 1848), p. 9.

Philadelphia opinion: "Insensibility During Surgical Operations Produced by Inhalation," *Medical Examiner and Record of Medical Science*, n.s., no. 24 (Dec. 1846): 719–20.

Date of first test in Philadelphia: "Operations in which ether or chloroform was used at the Clinic of the Jefferson Medical College," *Transcript of the American Medical Association,* vol. 1 (1848), appendix C–5, p. 221.

Darrach experiment: James E. Eckenhoff, *A History of Anesthesia at the University of Pennsylvania* (Philadelphia: Lippincott, 1966), p. 26.

Mott's trial: A. L. Cox, "Experiments with the Letheon in New York," *BMSJ* 35, no. 22 (Dec. 1846), p. 458; also, "Letheon in New York,"*BMSJ* 35, no. 21 (Dec. 1846), p. 142.

***Annalist*:** "The Letheon," *Annalist,* Jan. 15, 1847, p. 188.

Baltimore doctor's letter: "Insensibility Produced by Ethereal Inhalation," *BMSJ* 15, no. 22, (Dec. 30, 1846): 446.

Request for ether: H. Smith, Dec. 12, 1846, MGH, MC3, box 9, file 156.

Chapter 11: Repelled by a Common Moment

Jackson's version: *Manual of Etherization,* Charles T. Jackson, M.D. (Boston: Mansfield, 1861), pp. 18, 19–20,48

Chapter 12: Charles Jackson's Universe

Jackson childhood: Lydia Jackson to Lucy Brown, Oct. 5, 1821, in *The Selected Letters of Lidian Jackson Emerson,* edited by Delores Bird Carpenter (Columbia: University of Missouri, 1987), pp. 4, 12; also, Ellen Tucker Emerson, *The Life of Lidian Jackson Emerson,* edited by Delores Bird Carpenter (Boston: Twayne, 1980), pp. 3–13.

Father's advice: Charles Jackson [father of CTJ], *Book of Moral Advice to his Daughters,* handwritten booklet, Houghton, bMS 1280.235 (470).

Jackson at school: C. T. Jackson to C. Brown, Dec. 9, 1823, Houghton bMS Am 1280.226 (3684); also, Jan. 10, 1824, ibid. (3687).

Bypassing Harvard: C. T. Jackson to C. Brown, May 2, 1825, ibid. (3692).

Parisian studies: C. T. Jackson to Lidian Emerson, Nov. 18, 1832, Library of Congress, Manuscript Collection, Charles Jackson papers, MS 64-1424; also, C. T. Jackson to C. Brown, Jan. 28, 1831, Houghton bMS Am 1280.226 (3696).

The *Sully*: William T. Davis, *Plymouth Memories of an Octogenarian,* (Plymouth: Memorial Press, 1906), p. 274.

Remark about companion: C. T. Jackson to Lydian Emerson, Sept. 15, 1835, Houghton bMS Am 1280.226 (3705).

Jackson general account of voyage: C. T. Jackson to C. Brown, Nov. 17, 1832, ibid. (3697).

Morse recollections: *A Memorial of Samuel F. B. Morse from the City of Boston* (Boston: City Council, 1872), pp. 66, 72–76.

Jackson recollection: Ibid., pp. 75–76.

Difficulty of making living: C. Jackson to Lucy Brown, Jan. 15, 1835, Houghton bMS Am 1280.229 (17).

Maine trip: C. T. Jackson to Lidian J. Emerson, Nov. 7, 1837, Houghton bMS Am 1280.226 (3707).

Lydia meets RWE: Ellen Tucker Emerson, *The Life of Lidian Jackson Emerson,* ibid., p. 43.

Comments on Harvard: C.T. Jackson to H.A.S. Dearborn, Feb. 13, 1857, BPL.

Bartol anecdote: C. A. Bartol, "Charles Thomas Jackson," letter to the editor, *Boston Daily Advertiser,* Sept. 2, 1880, p. 3.

Fellow geologist's description: "Charles Thomas Jackson," *American Geologist,* J. B. Woodworth, 20–2, August 1897, p.83

Bouvé opinion: Prof. T. T. Bouvé, "Dr. Charles T. Jackson," *Proceedings of Boston Society of Natural History* 21 (Oct. 1880): 40–47.

Morse correspondence: *A Memorial of Samuel F.B. Morse,* ibid, p. 66,

Chapter 13: Horace Wells in Paris

Wells letter to the editor: reprinted in Erving, "The Discoverer of Anæsthesia," 426–47.

Interest in birds: Address by Charles C. Wells (son), in *Discovery of Anesthesia by Dr. Horace Wells,* p. 97.

Ellsworth letter: P. W. Ellsworth, "The Discoverer of the Effects of Sulphuric Ether," *BMSJ* 35, no. 20 (Dec. 1846): 397.

Ellsworth article: P. W. Ellsworth, M.D., "On the Modus Operandi of Medicines," *BMSJ* 17, no. 19 (Jun. 11, 1845): 396.

Lancet jump-up behinders: Hodges, *A Narrative of Events,* p. 118.

Paris correspondent on claimants: Dr. Brewster to *Journal of Commerce,* Mar. 26, 1847, reprinted in Archer, "Life and Letters of Horace Wells," 127.

Description of Jackson campaign: Ibid., 128.

Letter from South Reading doctor: J. S. Mansfield letter, *BMSJ,* Dec. 23, 1846, pp. 424–25.

Morton's letter to war secretaries: Edward J. Warren for Dr. Morton to Secretary of the Navy, Jan. 18, 1847, MGH, MC3, box 9, file 143.

Wells letter to his mother, about Paris: H. Wells to mother and sister, Mar. 28, 1847, reprinted in Archer, "Life and Letters of Horace Wells," *Journal of the American College of Dentists* 12, no. 2 (June 1945): 86.

William Beaumont diary: A. Scott Earle, ed., *Surgery in America* (Philadelphia: W.B. Saunders, 1965), p. 84.

Jackson at Lake Superior: "Charles Thomas Jackson," *American Geologist* 20, no. 2 (Aug. 1897): 79.

Wells on the high seas: Davis, *Plymouth Memories,* p. 275.

Edward J. Warren and Morton: "The Problem of the Three Edward Warrens," *New England Journal of Medicine* 224, no. 25 (June 1941): 1075.

Wells tries nitrous oxide in operations: Hodges, *A Narrative of Events,* pp. 97–105.

Chapter 14: Chlory

Carey letter: Milan G. Carey to D. L. Allen, Apr. 15, 1847, Countway, bMS misc. C.

Physician at meeting: *An Appeal to Patrons of Science,* (Boston: n.p., 1857), p. 59.

Clergyman on etherization: William H. Welch, "A Consideration of the Introduction of Surgical Anæsthesia"(Boston: Barta Press, n.d. [1908]), p. 21.

Negative aspects of ether: untitled lead article on ether, *Littell's Living Age* 161, June 12, 1847, pp. 491–92.

Simpson account of trials: Simpson letter of Dec. 4, 1847, reprinted in "Sir James Simpson's Introduction of Chloroform by his Daughter," *Century Magazine* 47, no. 3 (Jan. 1894), p. 415.

Queen Victoria: Virginia S. Thatcher, *History of Anesthesia* (New York: Garland Publishing Co., 1953), p. 16.

Simpson's motto: "Sir James Simpson's Introduction of Chloroform," p. 420.

Rabbit anecdote: "Remarks of Lord Playfair," *The Semi-Centennial of Anesthesia*, p. 83.

Hanging: "On the Use of Chloroform in Hanging," *American Whig Review*, Sept. 1848, p. 296.

Elephant story: "Desctruction of an Elephant," *Boston Evening Transcript*, Aug. 14, 1855, p. 2.

Bourbaugh on malingerers: Dr. Charles C. Bourbaugh, "Chloroform," *United States Service Magazine*, Apr. 1865, p. 327.

James Graham: *The Age of Agony*, Guy Williams (Chicago: Academy Chicago, 1986) p. 192

Richardson on ether drinkers: Benjamin Ward Richardson, "The Chloral and Other Narcotics," *Contemporary Review* 35, Apr.–Aug, 1879, p. 729.

Natural addictive substances: ibid., p. 723.

Detroit chloroform addict: "The Chloroform Habit as Described by One of its Victims," *Detroit Lancet* 8 (1884–85): 251–54; reprinted in H. Wayne Morgan, ed., *Yesterday's Addicts* (Norman: University of Oklahoma Press, 1974), pp. 147–52.

Connecticut legislature: "Resolution of the General Assembly of Connecticut of May, 1847 [Doctor Horace Wells]," reprinted in *Dr. Wells, The Discoverer of Anæsthesia*, p. 10.

Wells and Terry: "Dentist," advertisement in *Hartford Daily Courant*, Jan. 25–31, 1848.

Receipt for Wells auction: Archer, "Life and Letters of Horace Wells Discoverer of Anesthesia," 134–35.

CHAPTER 15: ALL ALONE AT THE TOMBS

***New York Herald* on women:** "Outrageous," *New York Herald,* January 23, 1848, p. 2.

Attacks: "Assault with Vitriol," *New York Daily Tribune,* January 24, 1848, p. 3.

Wells ad in paper: Archer, "Life and Letters of Horace Wells," 135–36.

Kerr on chloroform addict: Norman Kerr, *Inebriety: Its Etiology, Pathology, Treatment and Jurisprudence* (London: H.K. Lewis, 1888), pp. 104–5.

New York neighborhood: Hone, *The Diary of Philip Hone,* p. 396.

Wells account: "Distressing Case of Suicide," *New York Tribune,* Jan. 25, 1848, p. 2; also, *New York Herald,* Jan. 25, 1848, p. 2.

Covel version: J. C. Covel, letter to the editor, *New York Tribune,* Jan. 27, 1848, p. 2.

Criticism of medical help: "The Recent Suicide," *New York Tribune,* Jan. 26, 1848, p. 3.

Copycat suicide: "Suicide," *Hartford Daily Courant,* Jan. 29, 1848, p. 2.

***New Haven Journal*:** "The Suicide of Dr. Wells," reprinted in *Hartford Daily Courant,* Jan. 27, 1848, p. 2.

Policeman contacts accuser: "More About the Suicide," *New York Evening Post,* Jan. 27, 1848, p. 2.

CHAPTER 16: IN THE LOBBY OF THE WILLARD HOTEL

Gould on controversy: Hodges, *A Narrative of Events,* p. 61.

Morton's "Memoir": Morton, *A Memoir,* p. 3.

***Littell's* articles:** Richard H. Dana, Jr., "The Ether Discovery," *Littell's Living Age* 16, Mar. 18, 1848, pp. 529–71; Joseph L. Lord and Henry C. Lord, "The Ether Controversy," *Littell's Living Age* 17, June 10, 1848, pp. 491–522.

Lake Superior problem: RWE to William Emerson, May 23, 1849; also, "The New Geologists," *Boston Bee,* July 2, 1849

Jackson on Lake Superior problem: C. T. Jackson to G. C. Swallow, Apr. 22, 1849, Countway bMS misc. Jackson; also, C. T. Jackson to RWE, Apr. 24, 1849, Houghton, bMS AM 1280.

Morton telegram: W. T. G. Morton to N. I. Bowditch, telegram, MGH, MC3, box 1, file 7, item 1-n.

Long article: Crawford W. Long, "An Account of the First Use of Sulphuric Ether," *Southern Medical and Surgical Journal* 5 (1849), reprinted in The History of Anesthesiology reprint series no. 1 (Oak Ridge, Ill.: Wood Library-Museum, 1992), pp. 3–13.

William Clarke incident: Phyllis Allen Richmond, "Was William E. Clarke of Rochester the First American to Use Ether for Surgical Anesthesia?" *Rochester Historical Society Scrapbook* 1, 1950, pp. 11–12.

"Greek physician" remark: Ashhurst, "Surgery Before the Days of Anæsthesia," p. 37.

Academy of Sciences decision: "Report of the Premiums Awarded in the Departments of Medicine and Surgery, for the Years 1847 and 1848," reprinted in *Statements,* p. 373.

Jackson on the "Enemy:" C. T. Jackson to RWE, July 16, 1852, Houghton, bMS Am 1280 (1670).

Morton's entertaining: "Who Invented Anesthesia," *New York Times,* Nov. 15, 1858, p. 21.

Congress on Morton dossier: reprinted in *Statements,* p. 36.

Telegraph conversations: personal communication from Mrs. Charles H. Sheridan, William W. Ford, cited in "William Thomas Green Morton," *More Books: The Bulletin of the Boston Public Library,* Oct. 1846, p. 302, n. 22.

Mrs. Hale's article: Sarah Josepha Hale, "Etherton Cottage and the Discoverer of Etherization," *Godey's Lady's Book* 46, Mar. 1853, p. 212.

Morton factory: *Townsman* (Wellesley, Mass.), Dec. 25, 1906, Wellesley (Mass.) Free Library clipping file: Morton.

Eastern Railroad: Francis B. C. Bradlee, *The Eastern Railroad* (Salem, Mass.: Essex Institute, 1917), pp. 55–56.

CHAPTER 17: EMERSON'S WIFE'S BROTHER

Agassiz visit: Hodges, *A Narrative of Events*, pp. 97–98.

Bouvé recollection: T. T. Bouvé, "Sketch of Dr. Charles T. Jackson," *Popular Science Monthly* 19, July 1881, p. 406.

Bartol quote: C. A. Bartol, "Charles Thomas Jackson," letter, *Boston Daily Advertiser*, Sept. 2, 1880, p. 3.

Sanborn quotes: newspaper clipping attributed to "F.B.S.," "The Late Dr. Jackson of Boston," F.B.S., Sept. 4, 1880, MHS, C. T. Jackson papers, p. 491.

"That monomaniac": Amos Kendall, *Morse's Patent* (Washington, D.C.: Jno. T. Towers, 1852), p. 3.

Ralph Waldo Emerson on Jackson's mood: letter to [Louis] Agassiz, Jun. 13, 1859, Charles T. Jackson Papers, MHS.

Emerson's private opinion: RWE to Lidian Emerson, Apr. 20–21, 1848, in Ralph Rusk, ed., *The Letters of Ralph Waldo Emerson* (New York: Columbia University Press, 1939), vol. 4, p. 57.

Emerson on hospital: letter to John Adam Kasson, draft copy, Apr. 26, 1864, Charles T. Jackson Papers, MHS.

Emerson more blunt on hospital: draft of a letter to Charles Sumner, Jan. 31, 1872, ibid.

Andrews on Jackson's visit: C. H. Andrews to Edwin Newton, Mar. 22, 1900, reprinted in Joseph Jacobs, *Dr. Crawford W. Long* (Atlanta: [n.p.] 1919), p. 41.

Jackson on Morton in Washington: C. T. Jackson to RWE, Apr. 17, 1854, Houghton bMS AM 1280.

Wales's threat: letters to Charles T. Jackson, July 3, 1858, Charles T. Jackson Papers, MHS.

Chapter 18: Morton's Civil War

Tuckerman and director: "Trial of W.S. Tuckerman on Charge of Embezzlement—Second Day," unidentified clipping, MHS, C. T. Jackson papers.

Stockholders' meeting: "Adjourned Meeting of the Eastern Railroad Stockholders," *Boston Daily Times,* July 31, 1855, p. 4; *Boston Evening Transcript,* Dec. 31, 1856, p. 1; also, *Boston Evening Transcript,* July 30, 1855, p. 2; also assorted clippings MHS, C. T. Jackson papers; also *Boston Traveller,* Dec. 3, 1857, p. 1.

New York effort for Morton: "The Invention of Anæsthesia—National Testimonial," John Watson, letter to the editor, *New York Times,* Dec. 3, 1858, p. 2; also, Valentine Mott M.D. and Willard Parker M.D., letter to the editor, "Who Discovered Anæsthesia?—Statement of Drs. Mott, Francis and Parker," ibid., Dec. 20, 1858, p. 2.

Rice on the book: "A Grain of Wheat from a Bushel of Chaff," *The Knickerbocker,* 53–2, Feb. 1859, pp 137–38.

Morton's effigy: "The Wellesley Buttonwood," F. H. Gilson, *Our Town* (Wellesley, Mass.) 5, no. 12 (Dec. 1902): 141.

Lawsuit against Eye Infirmary: "The Patent for Sulphuric Ether," *New York Times,* Jan. 31, 1862, p. 6.

Union surgeon on Antietam: "Military Surgery Ancient & Modern," *United States Service Magazine,* Feb. 1864, p. 132.

Bartlett operation: "The Surgeon at the Field Hospital," *Atlantic Monthly* 46, Aug. 1880, pp. 185–86.

Thurston observations: H. Thurston to "Sir," Jan. 8, 1864, Countway, bMS c70.2.

AP article: Excerpted in *Biographical Sketch of Dr. William T. Morton,* booklet, New York Public Library.

Morton's need for PR: W. T. G. Morton to "Major," May 13, 1864, Countway, bMS misc.

Morton's booklet: W. T. G. Morton, *The Use of Ether as an Anesthetic* (Chicago: American Medical Association, 1904).

Congressman's observation: J. A. Kasson to RWE, Sept. 29, 1864, Countway, bMS c70.3.

Schoolteacher on Morton lecture: "Anæsthesia," *Harper's* 31, Sept. 1864, p. 453.

Chicago physician: "Wm. T.G. Morton and his Extraordinary Pretensions," *Chicago Medical Examiner* 6, no. 12 (Dec. 1865,): 743–51.

Pro-Jackson article: "The Discovery of Etherization," *Atlantic Monthly*, June 21, 1868, pp. 718–25.

CHAPTER 19: POISON IN THE BEAMS

***Mrs. Morton on* article:** Morton, "The Discovery of Anesthesia," p. 318.

***Harper's* article:** "Anæsthesia," p. 453.

New York in heat wave: "Close of the Heated Term—Marked Change in the Atmosphere," *New York Times,* July 18, 1868, p. 8.

Account of Morton's spell: "The Death of Professor Morton," *New York Herald,* July 17, 1868, p. 5; also, "Death of Prof. Morton," *New York Times,* July 17, 1868, p. 5; also "Sudden Death of Dr. Morton," *Boston Transcript,* July 17, 1868, p. 2.

Morton on list of victims: "Heavy Mortality in this City," *New York Post,* July 16, 1868, p. 2.

CHAPTER 20: DR. JACKSON'S SCRAPBOOK

Dr. Jackson's scrapbook: C. T. Jackson, Houghton Med 1945.8*.

Cohasset: Ellen T. Emerson letter to "Miss Waterman," July 13 [1865], Houghton: Ellen T. Emerson letter file, bMS Am 1280.235 (703), vol. 4.

Jackson's California trip: *Proceedings of the Boston Society of Natural History* 10 (Nov. 1865): 224–26.

Canadian Rockies trip: Ibid., 12 (June 1868): 88–89.

Financial problems: RWE to Abraham Jackson, Aug. 11, 1869, in Rusk, *Letters,* vol. 6, p. 81.

Turns down presidency of society: C. T. Jackson letter, *Proceedings of the Boston Society of Natural History* 14 (June 15, 1870): 2.

Holmes cites Jackson at party: Lidian J. Emerson to RWE, Feb. 1, 1867, Carpenter, *Selected Letters of Lidian Jackson Emerson*, p. 255.

Holmes private opinion: O. W. Holmes to R. Dana, letter excerpted in Eleanor M. Tilton, *Amiable Aristocrat* (New York: Henry Schuman, 1947), p. 188.

Ellen on Emerson's return to town: Ellen Tucker Emerson, *The Life of Lidian Jackson Emerson*, pp. 161–62.

Ellen on party preparations: Ellen Tucker Emerson to Aunt Edith, June 22, 1873, Houghton bMS Am 1280.235, box 29 [page missing from letter].

Jackson becomes ill: Ellen Tucker Emerson, *The Life of Lidian Jackson Emerson*, p. 164.

Initial reaction: Lidian Emerson to daughter, Ellen, July 29 and Aug. 6, 1873, Carpenter, *Selected Letters of Lidian Jackson Emerson*, pp. 310–11.

Wife's observation: Susan Jackson to RWE, Aug. 10 [1873], Houghton bMS 1280 (1678).

Care assumed by MGH: Thos. B. Hall to Susan Jackson, Nov. 28, 1873, Countway bMS c70.3.

Epilogue: Since Ether Day

Interval narcosis: Vincent J. Collins, *Principles of Anesthesiology* (Philadelphia: Lea and Febiger, 1976), p. 12.

Becomes a specialty: Thatcher, *History of Anesthesia*, pp. 23–25.

Use of nitrous oxide: Robert M. Julien, *Understanding Anesthesia* (Menlo Park, Calif.: Addison-Wesley, 1984), pp. 91–92.

Unknown process of anesthesia: Collins, *Principles of Anesthesiology*, p. 1288; also, H. P. Franks and W. R. Lieb, "Molecular and Cellular Mechanisms of General Anaesthesia," *Nature*, vol. 367, Feb. 1994, p. 607.

Bibra and Harless's theory: John Adriani, *The Chemistry and Physics of Anesthesia* (Springfield, Ill.: Charles Thomas, 1979), p. 561.

Collins quote: Collins, *Principles of Anesthesiology*, p. 1292.

SELECTED BIBLIOGRAPHY

BOOKS AND MAJOR ARTICLES

Archer, W. Harry. *Chronological History of Horace Wells*. Pittsburgh, Penn.: A.R. Plantz, 1939.

———. "Life and Letters of Horace Wells: Discoverer of Anesthesia," *Journal of the American College of Dentists* 11, no. 2 (June 1944): 83–197; continued in 12, no. 2 (June 1945): 85–99.

Fitz, Reginald. "The Value of Imponderables." *New England Journal of Medicine* 236, no. 16 (Apr. 1947), pp. 555–62.

Hodges, Richard M. *A Narrative of Events Connected with the Introduction of Sulphuric Ether into Surgical Use* (Boston: Little, Brown, 1891).

Jackson, Charles T., M.D. *Manual of Etherization*. Boston: Mansfield, 1861.

Jacobs, Joseph. *Dr. Crawford W. Long*. Atlanta, 1919; reprint, Athens, Ga.: Crawford Long Museum, n.d.

Kendall, Amos. *Morse's Patent: Full Exposure of Dr. Chas. T. Jackson's Pretensions*. Washington, D.C.: Towers, 1852.

Morton, W. T. G. *A Memoir to the Academy of Sciences at Paris on a New Use of Sulphuric Ether*. New York: Henry Schuman, 1946.

Nagle, David R. "Anesthetic Addiction and Drunkenness: A Contemporary and Historical Survey. "*International Journal of the Addictions* 3, no. 1 (Spring 1968): 25–39.

Slade, Daniel Denison. "The First Capital Operation Under the Influence of Ether." *Scribner's Magazine*, July 1892, p. 518.

Smith, W. D. A. *Under the Influence*. Park Ridge, Ill.: Wood Library-Museum of Anesthesiology, 1982.

Teller, Barbara Gorely. "William Thomas Green Morton: Scientific Farmer," *Wellesley Townsman*, Dec. 19, 1981; Jan. 12, 1982.

Warren, J. Collins. "The Influence of Anesthesia on the Surgery of the 19th Century: An address Before the American Surgical Association," privately printed. Boston, 1906.

Wolfe, Richard J. *Robert C. Hinckley and the Recreation of the First Operation Under Ether*. Boston: Boston Medical Library in the Francis A. Countway Library of Medicine, 1993.

Woodworth, J. B. "Charles Thomas Jackson." *American Geologist*, Aug. 1897, with a bibliography of his writings.

ACKNOWLEDGMENTS

Ether Day, October 16, 1846, is still recognized in Boston, and especially at Massachusetts General Hospital. Jeff Mifflin, the archivist and curator there, was this book's first friend. At those times when the research ground to a halt, he was always resourceful in setting things right again. I am also grateful to have had the opportunity to work with Betty Falsey at Houghton Library of Harvard University and Jack Eckert of the Boston Medical Library at Countway, of the Harvard Medical School.

The search for pictures of Horace Wells kept on leading to the articles of Dr. Harry Archer, a professor at the University of Pittsburgh School of Dentistry. The articles had the most intriguing pictures seen anywhere, but unfortunately they dated from the mid-1940s. Dr. Stephen Kondis, of the School of Dentistry, found time in the middle of a busy schedule to search for Dr. Archer's original glass slides. He also braved a snowstorm to have them turned into prints. I thank Dr. Kondis for the three Wells pictures in this book, and for an adventure that unfolded day by day at the other end of a telephone for me, with a happy ending. Patrick Sim and Karen Bieterman at the Wood Library-Museum of Anesthesiology, John Pollack at the University of Pennsylvania, Michelle Marcella at Massachusetts General Hospital, and Laurel Nilsen at the Wellesley Historical Society also went to extra trouble for the sake of the illustrations herein.

Tim Duggan of HarperCollins, and Karl Weber, my agent, shaped this book with their suggestions. Susan Llewellyn, who edited the text, deserves very special thanks. I would also like to thank my mother and father for willingly conversing about ether for the past five years—and for giving me a place to stay in Boston. Paul M. Birchmeyer helped me. And always, Neddy.

INDEX

Page numbers in *italics* refer to illustrations.